Alexander Kukushkin

Radio Wave Propagation in the Marine Boundary Layer

Alexander Kukushkin

Radio Wave Propagation in the Marine Boundary Layer

WILEY-VCH

WILEY-VCH Verlag GmbH & Co. KGaA

Editors

Alexander Kukushkin
Gordon, NSW 2072, Australia
e-mail: alexk_24@yahoo.com.au

Cover Picture
The image on the cover is from NASA's 2.84 GHz Space Range Radar (SPANDAR) at Wallops Island, Virginia. The image corresponds to a ducting event on April 2, 1998. Image courtesy of Space and Naval Warfare Systems Center, San Diego, CA.

Library of Congress Card No.:
applied for

British Library Cataloguing-in-Publication Data
A catalogue record for this book is available from the British Library.

Bibliographic information published by Die Deutsche Bibliothek
Die Deutsche Bibliothek lists this publication in the Deutsche Nationalbibliografie; detailed bibliographic data is available in the Internet at <http://dnb.ddb.de>.

Printed in the Federal Republic of Germany.

Printed on acid-free paper.

Typesetting Kühn & Weyh, Satz und Medien, Freiburg
Printing betz-druck GmbH, Darmstadt
Bookbinding Großbuchbinderei J. Schäffer GmbH & Co. KG, Grünstadt

ISBN 3-527-40458-9

Preface

This book is about the parabolic approximation to a diffraction problem over a sea surface. While the parabolic equation method in radio wave propagation over the earths surface was introduced by V.A. Fok almost fifty years ago, its popularity has grown recently due to the development of advanced computational methods based on the parabolic approximation. Numerous computational techniques have been evolved and used for analysis of radio- and acoustic wave propagation in either deterministic or random media.

This book is concerned with the analytical solution to a problem of wave propagation over the sea surface in the atmospheric boundary layer. Two basic mathematical methods have been used, depending on the ease of obtaining a closed analytical solution:

1. Expansion of the quantum-mechanical amplitude of the transition into a complete and orthogonal set of eigen functions of the continuous spectrum.
2. The Feynman path integral.

It is not intended to provide a full step by step mathematical background to the above methods but, rather, is dedicated to the application and analysis of the physical mechanisms associated with the combined effect of scattering, diffraction and refraction. The mathematical foundations for the above methods can be found in numerous monographs and handbooks dedicated to quantum mechanics and mathematical theory.

The book is arranged as follows: Chapter 1 presents the basic assumptions used to describe the propagation media, i.e. the atmospheric boundary layer. It provides a simplified description of the turbulent structure of the refractive index in the atmospheric boundary layer and summarises the model of the troposphere to be used in the analysis of the wave propagation. It introduces some foundation for the composition of the refractive index as two components: a deterministic layered structure and a relatively small-scale random component of turbulent refractive index. A basic classification of the propagation mechanisms, such as refraction, ducting, diffraction and scattering is briefly introduced according to the presence and value of the negative gradients of refractivity in the troposphere.

Chapter 2 commences with an overview of the mathematical methods developed for analysis of the problem of wave propagation and scattering in a stratified medi-

Radio Wave. Alexander Kukushkin

ISBN: 3-527-40458-9

um with random fluctuations of the refractive index. It also positions the method introduced in this book as an extension of the well-known analogy between the quantum-mechanical problem of the quasi-stationary states of the Schrödinger equation and the problem of radio wave propagation in the earths troposphere. The advantage of using this approach is that the Green function to the parabolic equation is expanded over the complete set of orthogonal eigen functions of the continuous spectrum. This representation is equivalent to a Feynman path integral which is used in Chapter 3 to investigate the higher order moments of the wave field over the surface with impedance boundary conditions.

Some new physical mechanisms associated with scattering are analysed and explained in Chapter 3.

Chapter 4 introduces a perturbation theory for normal waves in a stratified troposphere. The problem here is that the common perturbation theory does not work for equations with a potential unlimited at infinity. Such potentials appear in the problem of an electron in a magnetic field or in radiowave propagation over the earths surface in the parabolic equation approximation. A modified perturbation theory is applied to the analysis of the spectrum of normal waves (propagation constants) for the boundary problem with a somewhat arbitrary profile of the refractive index. The analytical solution and numerical results are discussed for two practically important models of refractive index in the near-surface domain: the bilinear approximation and the logarithmic profile. Also in Chapter 4, we present a closed analytical solution for a second moment of the wave field (coherence function) in the presence of an evaporation duct filled with random inhomogeneities of refractive index. The mechanism of interaction between discrete and continuous modes due to scattering of the random irregularrities in the refractive index is analysed in detail.

Chapter 5 deals with the elevated tropospheric duct. We start from a normal mode structure for the trilinear profile of the refractive index and analyse the wave field in geometric optic approximation thus introducing rays and modes. The specific case of the presence of two waveguides, elevated duct and evaporation duct, simultaneously, is analysed in detail by means of presenting the mechanism of exchange of the wave field energy in a two-channel system.

In Chapter 5 we also introduce the mechanism of excitation of the normal waves in an elevated duct by means of single scattering on turbulent irregularities of refractive index. This case may represent significant practical interest in the case of ground–air communication for two reasons: first, the elevated ducts are often detached from the surface and the near-surface antenna is ineffective in excitation of the trapped modes, and secondly, strong anisotropic irregularities are often present in the upper boundary of the elevated tropospheric duct due to the physics of its creation and, therefore, can produce a significant scattering effect of the incoming waves from a surface-based antenna.

Finally, in Chapter 6 we analyse some non-conventional mechanisms of the overhorizon propagation. First, the effect of a stochastic waveguide created by anisotropic irregularities in the refractive index. This mechanism is analysed in terms of the perturbation theory presented in Chapter 3. The second mechanism is a single scattering of diffracted field in the earths troposphere. This mechanism is rather complementary to a conventional single-scattering theory; it cannot explain the observed levels of the signal but, contrary to conventional theory, reveals a correct behaviour with regard to frequency.

The Appendix provides a brief theory of the Airy functions and some asymptotic representations.

The analytical solutions and results considered in this book are chiefly applicable to radio propagation in the UHF/SHF band, i.e., from 300 MHz to 20 GHz where refraction and scattering play a major role in anomalous propagation phenomena, such as a waveguide mechanism in tropospheric ducts.

I want to thank my former colleagues I. Fuks and V. Freilikher in cooperation with whom most of the theoretical studies have been performed. It was my privilege to work with my colleagues in such a productive and encouraging environment. I wish to thank V. Sinitsin who introduced me to research activities in this area and supported me at the start of my career.

Most of all, I am gratefully obliged to my lovely wife, Galina, for making it possible.

Alexander Kukushkin
Sydney, Australia, March 2004

Contents

Radio Wave. Alexander Kukushkin

ISBN: 3-527-40458-9

1
Atmospheric Boundary Layer and Basics of the Propagation Mechanisms

The troposphere is the lowest region of the atmosphere, about 6 km high at the poles and about 18 km high at the equator. In this book we study radio wave propagation along the ocean and can reasonably assume that all processes of propagation occur in a lower region of the earth's troposphere. That lower region and the atmospheric conditions are of most importance for the subject under study here.

From the perspective of radio communications/propagation we limit our objective to an investigation of the impact of the atmospheric structure on the characteristics of the radio signal propagating through the atmospheric turbulence. All known methods of solutions to a similar problem are based on the separation of the space–time scales of the variations in both the refractive index n and electromagnetic field $\vec{E}$, $\vec{H}$ in two domains, described in terms of deterministic and stochastic methods. Intuition suggests that the spectrum of turbulent variations in the refractive index n will have the energy of its fluctuations confined to a limited space–time domain or, at least, have a clear minimum and, desirably, a gap spread over a significant interval in the time–space domain. It is apparent that the horizontal scales of variations in refractive index larger than the length of the radio propagation path have no immediate effect on the characteristics of the radio signal and rather affect its long term variations over the permanent path. This comes down to an upper boundary of the spatial variations of refractive index in a horizontal plane of about 100 km. The vertical irregularities are of most importance since they are responsible for the refraction and scattering of the radio waves in troposphere. However, there are some natural limitations on the region of the troposphere which might be of interest in its impact on radio propagation. The troposphere is naturally divided into two regions: the lower part of the troposphere, commonly called an atmospheric boundary layer, and the area above, called clear atmosphere.

The electric properties of the troposphere can be characterised by the dielectric permittivity ε or the refractive index $n = \sqrt{\varepsilon}$. The numerical value of the non-dimensional parameter n is pretty close to unity, however even a relatively small deviation of the refractive index from unity may have significant impact on radio wave propagation. Therefore, common practice is to use another definition of the refractive index $N = (n-1) \times 10^6$ instead of n, measurable in so-called N-units. The refractive index N, also called the refractivity, has the following relationships with atmospheric pressure p, temperature T and humidity, the mass-fraction of the water vapor, q, in the air:

Radio Wave. Alexander Kukushkin

ISBN: 3-527-40458-9

$$N = \frac{A_N \cdot p}{T}\left(1 + B_N \frac{q}{T}\right) \tag{1.1}$$

where $A_N = 77.6$ N-units × K hPa^{-1}, $B_N = 7733$ K^{-1}. The components p, T and q are random functions of the coordinates and time. The stochastic behavior of the meteorological parameters p, T, q and, therefore, the refractive index N is caused by atmospheric turbulence.

There are several reasons to separate the region of the first 1–2 km of atmosphere over the earth's surface, called the atmospheric boundary layer, ABL. The upper boundary of the ABL is seen as the height at which the atmospheric wind changes direction due to a combined effect of the friction and Carioles force. Among those reasons are:

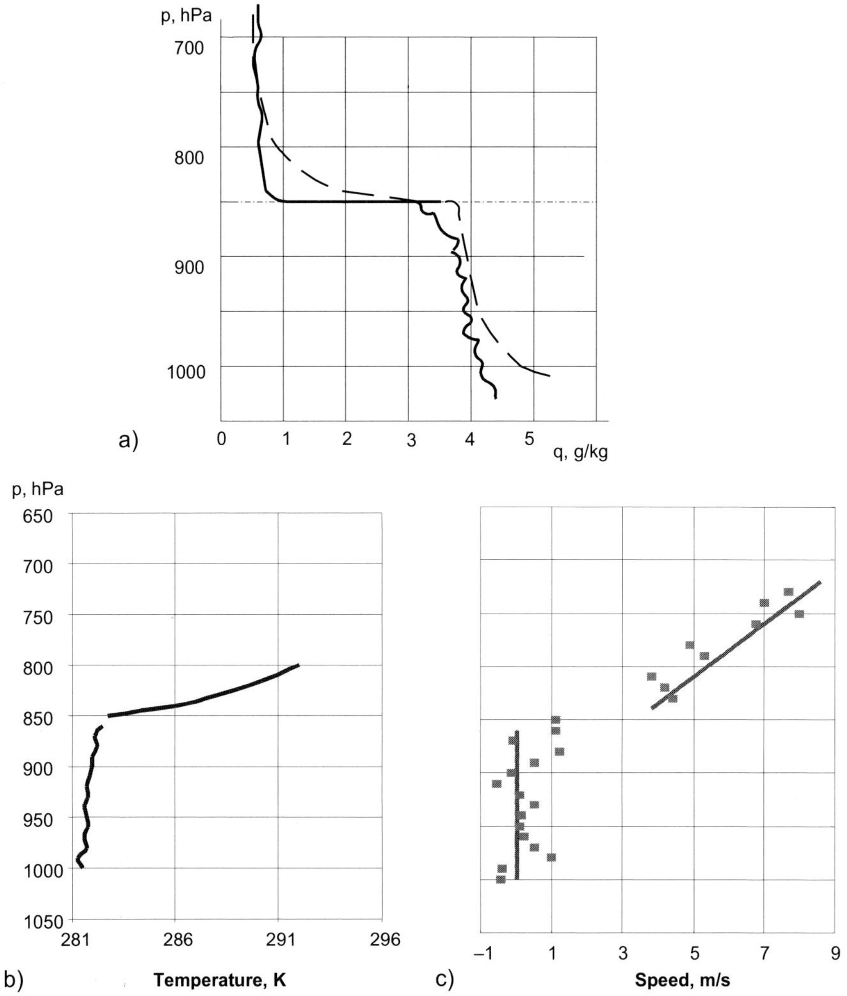

Figure 1.1 Meteorological parameters in the atmospheric boundary layer as function of height (atmospheric pressure, p): Humidity (a), temperature (b) and wind speed (c). All parameters experience sharp variations at the upper boundary of the atmospheric boundary layer.

- The interaction of the earth's surface and atmosphere is especially pronounced in this region.
- The meteorological parameters such as temperature, humidity and wind speed experience daily variations in this region due to apparent cyclic variations in the sun's radiation due to the earth's rotation.
- The ABL can be regarded as an area constantly filled with atmospheric turbulence. This is quite opposite to the atmospheric layer above the ABL, the so-called region of clear atmosphere, where turbulence is present only in isolated spots.
- The border between the clear atmosphere and the ABL is clearly pronounced with sharp variations in all meteorological parameters, as illustrated in Figure 1.1.

The spectrum of turbulent fluctuations in the atmospheric boundary layer is extremely wide: the linear scales of the variations range from a few millimetres to the size of the earth's equator, the time scales from tens of milliseconds to one year. Studies of the energy spectrum of the turbulent fluctuations of the meteorological parameters (temperature, humidity, pressure and wind speed) [1] have shown that the energy spectrum reveals three distinct regions: large scale quasi-two-dimensional fluctuations in a range of frequencies from 10^{-6}–10^{-4} Hz, the meso-meteorological minimum with low intensity of the fluctuations in the range 10^{-3}–10^{-4} Hz and a small scale three-dimensional fluctuation region with frequencies above 10^{-3} Hz.

Figure 1.2 shows the energy spectrum of fluctuations of the horizontal component of the wind speed in the atmosphere, taken from Ref.[1], where the ordinate corresponds to the product of the spectrum density $S(\omega)$ and the cyclic frequency ω of the variations in one of the meteorological components, and the abscissa corre-

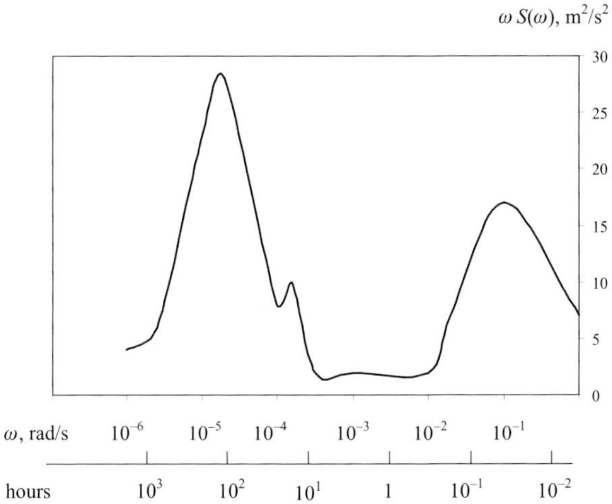

Figure 1.2 Energy spectrum of the fluctuations in the wind speed in the atmospheric boundary layer.

sponds to the frequency $\omega = 2\pi/T$, T being a period of variations. As observed in Figure 1.2, there are two major extremes of the function $\omega \cdot S(\omega)$: the high frequency maximum corresponds to a linear scale of turbulence of the order of tens and hundreds of meters, the low frequency maximum has a time scale of 5–10 days which is caused by synoptic variations (cyclones and anti-cyclones), the respective horizontal scale is thousands of kilometres and the vertical scale is of the order of 10 km. There is also an extended minimum in the spectrum $\omega \cdot S(\omega)$ that corresponds to the fluctuations with respective horizontal scales from 1 to 500 km and is called the meso-pause. The region of low frequency variation is called the macro-range while the region to the left of the meso-pause (high frequency variations) is called the micro-range.

The nature of the atmospheric turbulence is different in these two regions: in the macro-range the synoptic processes can be regarded as two-dimensional variations, while in the micro-range, with scales up to a hundred meters, the turbulence is three-dimensional and locally uniform. The mezo-pause is a transition region where a combined mechanism is observed. It is important to notice that by describing small-scale three-dimensional fluctuations in the micro-range region one can use Taylor's hypothesis of "frozen turbulence" which allows a transformation from time- to space-fluctuation scales by means of $L = 2\pi v/f$, where L is the spatial scale of the irregularities, f is the frequency of time variations in the refractive index, and v is the mean speed of the incident flow.

The basic conclusion that follows from the above observations is that, to some extent, the refractive index N and the dielectric permittivity ε can be presented as a sum of a slow varying component $\varepsilon_0(\vec{r})$ regarded as a quasi-deterministic function of the coordinates $\vec{r} = \{x, y, z\}$ and the random component $\delta\varepsilon(\vec{r})$. As observed from Figure 1.2, the quasi-deterministic component $\varepsilon_0(\vec{r})$ still varies in the horizontal plane and the energy of variations in the meso-pause minimum is not negligible. However, these variations have less impact on radio wave propagation than either variations of $\delta\varepsilon(\vec{r})$ in the micro-range or over-the-height variations in the "deterministic" component which may be responsible for a ducting in the troposphere.

Mathematically, the problem of radio wave propagation in a randomly inhomogeneous medium comes down to solving a stochastic wave equation with dielectric permittivity $\varepsilon(\vec{r}, t)$ which is a random function of coordinates and time. In many cases the process of propagation of a monochromatic wave in the troposphere can be considered in a quasi-steady state approximation, i.e. "frozen" in time. It is then convenient to represent the dielectric permittivity in the form $\varepsilon_0(\vec{r}) \equiv \varepsilon_0(z) + \delta\varepsilon(\vec{r})$, where $\varepsilon_0(z) = \langle \varepsilon_0(\vec{r}) \rangle$. The angular brackets denote averaging over the ensemble of the realisations of $\varepsilon(\vec{r})$. In fact, the mean characteristic of the tropospheric dielectric permittivity $\langle \varepsilon(\vec{r}) \rangle$ is commonly understood as a large-scale structure homogeneous in the horizontal plane and practically non-varying over the time over which the signal measurements have been performed and then, as a result of mathematical idealisation, $\langle \varepsilon(\vec{r}) \rangle = \varepsilon_0(z)$. In radio-meteorology mean characteristics of the meteo-parameters are usually understood to be the values obtained by averaging over a 30 min interval [2], i.e. averaging is performed over a frequency interval the lower limit of which is positioned within the limits of the meso-meteorological minimum.

The average characteristic obtained in this way is usually a function of the height z above the surface (sea, ground) and varies slowly with the horizontal coordinates and time. Assuming ergodicity and Taylor's hypothesis, such averaging over a time interval corresponds to the averaging over the ensemble of the realisations of $\varepsilon(\vec{r}, t)$.

As a compromise in analytical studies of radio wave propagation, a common approach is to neglect the residual variations in $\langle\varepsilon(\vec{r})\rangle$ over the horizontal coordinates, i.e. to regard the $\varepsilon_0(\vec{r}) \equiv \varepsilon_0(z)$. This assumption results in the introduction of the traditional model of a stratified atmosphere, in which the average values of refractive index N vary over the vertical coordinate z, the height above the ground. This traditional model provides some basis for a classification of the radio wave propagation mechanisms, in particular, a separation of the propagation into two classes: standard and non-standard. The following Sections 1.2 and 1.3 provide a brief analysis of the standard and non-standard models, while Section 1.4 deals with a statistical model for a random component of the refractive index.

1.1 Standard Model of the Troposphere

The standard mechanism of radio wave propagation is classified under the condition where the average vertical gradients of the refractive index $\gamma_N = dN/dz$ are close to the value $\gamma_N^{st} = -39$ N-units km^{-1}. Such conditions of refractivity constitute a model of *standard linear atmosphere* defined as

$$N = 289 - 39 \cdot z \tag{1.2}$$

and are applicable for heights less than 2 km.

Let us consider this model in detail involving a geometrical optic presentation for wave propagation.

Let us define "ray" as a normal to a wavefront propagating through the medium with varying refractivity $n(z)$. As is known, the ray bends in such a medium and the bending is defined by Snell's law. Introducing the horizontally stratified medium in terms of the set of thin layers with value of refractivity n_i, $i = 0, 1, \ldots$, such as illustrated in Figure 1.3, Snell's law can be written as

$$n_i \cos\phi_i = const \tag{1.3}$$

where ϕ_i is a sliding angle. Introducing the differentials of the ray direction dz, dS in the ith layer and differentiating both sides of Eq. (1.3) with respect to S:

$$\cos\phi_i \frac{dn_i}{dS} - n_i \sin\phi_i \frac{d\phi_i}{dS} = 0. \tag{1.4}$$

Substituting $dS = -dz/\sin\phi_i$, we obtain

$$\frac{d\phi_i}{dS} = -\frac{\cos\phi_i (dn/dz)}{n_i}. \tag{1.5}$$

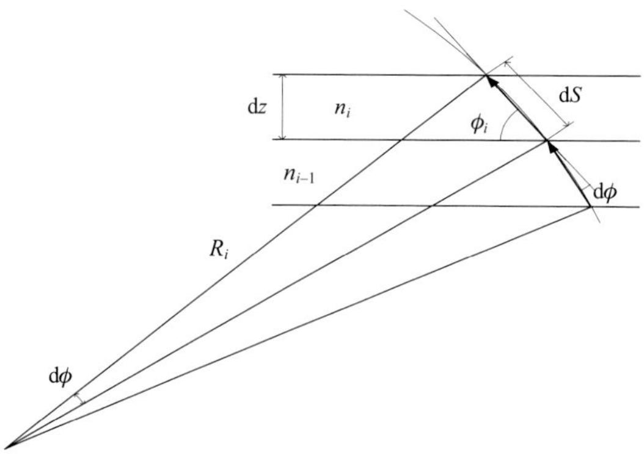

Figure 1.3 Illustration of Snell's law.

The radius of curvature at any point, $R_i = dS/d\phi_i$, and using Eq. (1.5) it results that

$$R_i = \frac{n_i}{\cos\phi_i} \frac{1}{(-dn/dz)}. \tag{1.6}$$

For the standard atmosphere with $dn/dz = 39 \times 10^{-6}\,\text{km}^{-1}$, the radius of curvature is given by

$$R_i = 25,000 \frac{n_i}{\cos\phi_i}. \tag{1.7}$$

If the launch angle ϕ_i is close to the horizontal, the ratio $\frac{n_i}{\cos\phi_i} \approx 1$ and a propagation path can be described as a circle of radius $R = 25{,}000$ km, Figure 1.4(a). By comparison, the radius of the earth's curvature is $a = 6370$ km, and $1/a = 157 \times 10^{-6}\,\text{km}^{-1}$. When the curvatures of both the propagation path and the earth are reduced by 39×10^{-6}, as in Figure 1.4(b), the propagation path has an effective curvature of zero (which is a straight line) and that hypothetical earth has an effective curvature of $(157 - 39) \times 10^{-6}\,\text{km}^{-1} = 118 \times 10^{-6}\,\text{km}^{-1}$. The equivalent radius of the sphere a_e can therefore be defined as

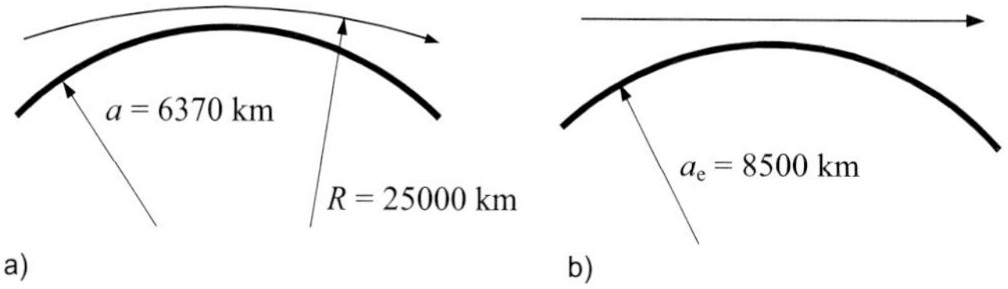

Figure 1.4 Introduction of the "effective" radius of the earth: a) Ray refraction in a "normal" atmosphere with "true" earth radius $a = 6370$ km. b) Effective ray refraction in case when the difference in curvatures of both ray and earth surface in (a) is compensated by introduction of the modified earth radius $a_e = 8500$ km.

$$\frac{1}{a_e} = \frac{1}{a} - \frac{1}{R} = \frac{1}{a} - |\gamma_N| \cdot 10^{-6} = \left(\frac{1}{118}\right) \times 10^6 \text{ km} = 8500 \text{ km}, \qquad (1.8)$$

approximately, $a_e = 4/3a$.

The modified refractive index can be defined as follows:

$$n_m(z) = n(z) + 1/a. \qquad (1.9)$$

It is apparent that the second term in Eq. (1.9) is a compensation for the earth's curvature. The modified refractivity $M(z)$ is then given by

$$M(z) = (n_m(z) - 1) \times 10^6 = N(z) + 0.157z. \qquad (1.10)$$

As observed from Eq. (1.10) for the case of standard refraction, when $\gamma_N = dN/dz = -39$ N-units km^{-1}, $\gamma_M = dM/dz = 118$ N-units km^{-1}. As explained in Section 2.1, modified refractivity comes logically from the parabolic approximation to the wave equations and effectively results in the introduction of the flat earth and additional curvature of the "rays" associated with radio waves propagating at low-angles along the earth's surface, as shown schematically in Figure 1.5.

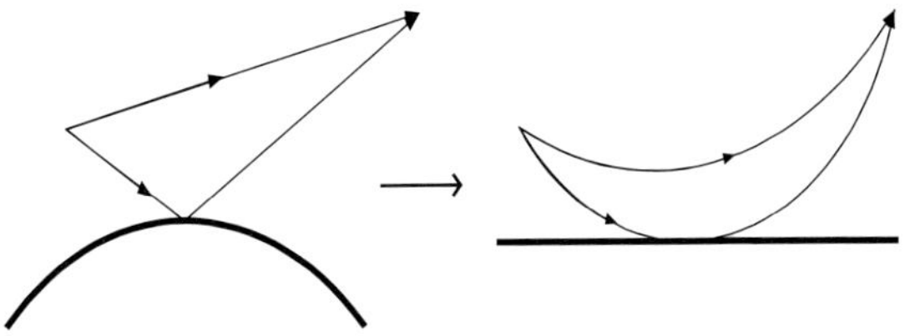

Figure 1.5 Impact from modified refractivity on the ray trajectories from a point source.

This traditional linear model of the troposphere provides some means of classification of the propagation mechanisms based on the value of the gradient of the refractivity in the lower troposphere, both conventional $\gamma_N = dN/dz$ and modified refractivity $\gamma_M = dM/dz$. A very broad classification is concerned with the separation of the propagation into two classes: standard and non-standard. Figure 1.6 represents schematically the ray traces for three basic sub-classes of standard mechanisms of propagation in the troposphere:

- Standard refraction, when the vertical gradient of refractivity is very close to $\gamma_N^s = -39$ km^{-1}, $\gamma_M^s = 118$ km^{-1}.
- Sub-refraction, when $\gamma_N > -39$ *N*-units km^{-1}, $\gamma_M > 118$ *N*-units km^{-1}.
- Super-refraction, when $\gamma_N < -39$ *N*-units km^{-1}, $\gamma_M < 118$ *N*-units km^{-1}.

As seen from Figure 1.6 the sub-refraction and super-refraction are associated with bending the waves outwards and inwards to the earth's surface respectively.

The standard propagation conditions can be characterised by the presence of two regions: the so-called line-of-sight (LOS) region, where the radio signal in the receiver has contributions from the direct wave and the wave reflected from the earth's

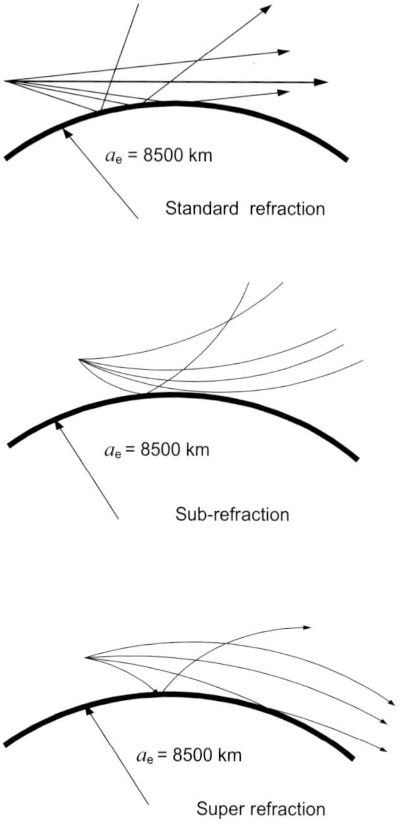

Figure 1.6 Classification of the propagation mechanisms.

surface; and the shadow region where the radio signal arrives as a result of diffraction over the earth surface. These two regions are separated at the range of horizon which can be defined as $\sqrt{2a_e} \cdot \sqrt{z}$, where the parameter z is the height of the observation point above the earth's surface. It should be noted that for a standard refraction $\sqrt{2a_e} = 4.12 \times 10^3\ \mathrm{m}^{1/2}$ compared with the value for optics of $3.83 \times 10^3\ \mathrm{m}^{1/2}$. For frequencies above 20 GHz the standard mechanisms include attenuation of the radio waves due to absorption by atmospheric gases (water vapor and oxygen).

Non-standard mechanisms of radio wave propagation are related to some anomalies in the vertical distribution of the refractive index, namely the gradients γ_M become negative. These anomalies are in turn caused by some extremes in the vertical profiles of either the humidity or temperature.

Under such conditions the waves can be trapped inside a localised region associated with such negative gradients of modified refractivity, this effect is called "ducting" and the radio wave propagation mechanism is somewhat similar to propagation in a waveguide.

1.2
Non-standard Mechanisms of Propagation

The term M-inversion is used to characterize the area of negative gradients of the M-profile, where $dM/dz < 0$. When M-inversions exist over some height interval, a radio-waveguide can be formed. The upper boundary of the waveguide is the height H at which $dM/dz = 0$, while the lower boundary z_{min} is given by $M(z_{min}) = M(H)$, see Figure 1.9 later. The waveguide is elevated if $z_{min} > 0$. If $z_{min} = 0$ or the equation $M(z_{min}) = M(H)$ has no solution, then the waveguide is classified as a surface waveguide, Figure 1.7.

Most often M-inversions are formed in the lower and upper regions of the atmospheric boundary layer because of the local interaction between the turbulent air mass and either the surface or the free atmosphere above the ABL [3].

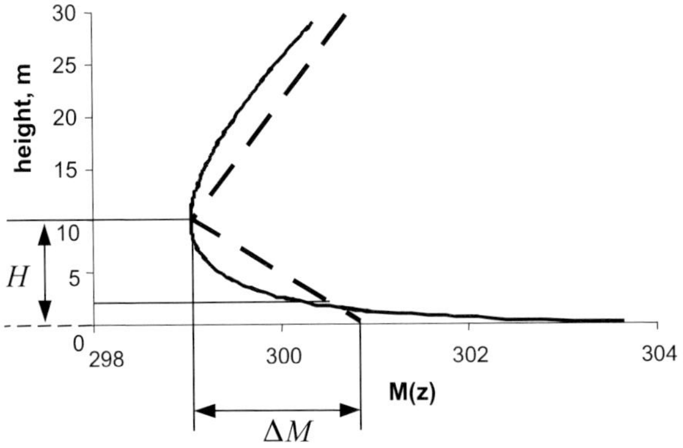

Figure 1.7 Evaporation duct, M-profile.

1.2.1
Evaporation Duct

One of the most important types of radio waveguide is the evaporation duct above the water surface [3] which has a nearly permanent presence in the subtropical regions of the Pacific Ocean. It is formed as a result of the air in contact with the sea surface becoming saturated by water vapor at the surface temperature. The water vapor content decreases logarithmically with height and the resulting M-profile has the form [3]:

$$M(z) = N(z_r) - H\alpha \ln\left(\frac{z}{z_r}\right) + \alpha z \tag{1.11}$$

where z_r is the roughness parameter of the sea surface and $\alpha = 10^6/a = 0.157\ \mathrm{m}^{-1}$ and $a = 6370$ km, the earth's radius.

The M-profile (1.11) is can be constructed from the data of a standard hydrometeorological measurement. It describes quite well the height dependence of the

M-profile under stable and neutral conditions of stratification in turbulence. The studies performed in Ref. [3] show that the thickness of the duct (inversion height H) is a function of the stability parameter of the turbulence at the water level layer and is functionally related to the inversion depth $\Delta M = M(z_r) - M(H)$, Figure 1.7.

It should be noted that no analytical solution to the wave equation exists for the M-profile in form (1.11). While a numerical solution can be adopted with some modifications to Eq. (1.11), in some applications it might be preferable to replace Eq. (1.11) by a profile for which the analytical solution to a wave equation is known. In particular, a common approach is to use a piecewise linear M-profile given by

$$M(z) = \begin{cases} M(H) + \alpha(z - H), & z > H \\ M(H) + \Delta M(H - z)/(H - z_r), & z \leq H \end{cases} \tag{1.12}$$

Comparison of the results obtained with both the piecewise linear approximation and the continuous M-profile (1.12) is presented in Section 4.1. It has shown a satisfactory agreement of the results of calculations for the lower order modes. In fact, the discrepancies observed for the high order modes in terms of their propagation constants and therefore the height functions are not really important because those modes are significant in the transition area between the shadow and LOS regions, where, in turn, the mode presentation is hard to utilise because of the very slow convergence of the modal series. On the other hand, it was observed that the errors introduced by replacement of the continuous profile by a piecewise approximation for the lower modes do not exceed the errors produced by uncertainties in determining the values of the refractive index from data of standard hydro-meteorological measurements. The overall conclusion therefore is that the approximation (1.12) is a valid practical approach to analytically evaluate the propagation mechanism in the evaporation duct. In computer based prediction systems, where input radio-meteorological information can be collected in nearly real-time with substantial precision, the hydro-dynamical model of the evaporation duct Eq. (1.11) can be easily implemented. The common problem of calculating the propagation constants of the trapped (or significant) modes can be reduced to pre-calculated look-up tables and interpolation.

Figure 1.8 shows a schematic ray bending in the evaporation duct, which allows over-the-horizon radio wave propagation.

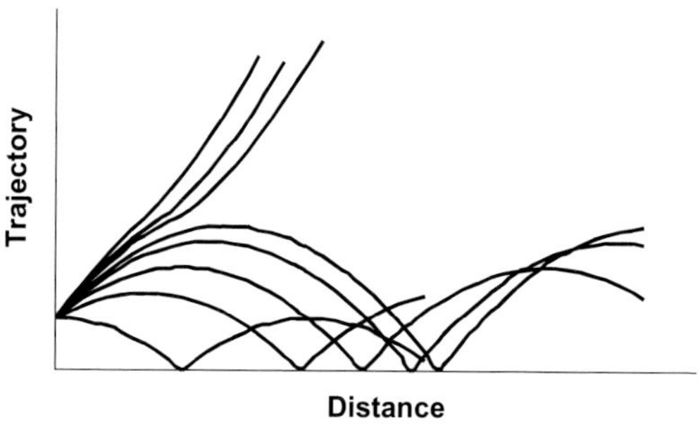

Figure 1.8 Ray trajectories in evaporation duct.

1.2.2
Elevated M-inversion

The most frequent cause of abnormally high signal level far beyond the horizon (200–500 km) is associated with the presence of elevated M-inversions. Those are formed in the subtropical region above the ocean as a result of powerful motion downwards in the center of an anticyclone and are usually called depression layers. The elevated M-inversions, also called elevated waveguides or inversion layers, are characterized by lower height of inversion H_i, the thickness of the layer with negative M-profile (M-inversion) $\Delta H = H_u - H_i$ and the depth of inversion, the so-called M-deficit $\Delta M_i = M(H_i) - M(H_u)$, as shown in Figure 1.9. Figure 1.10 shows schematic ray trajectories in an elevated duct when the source is placed inside its boundaries.

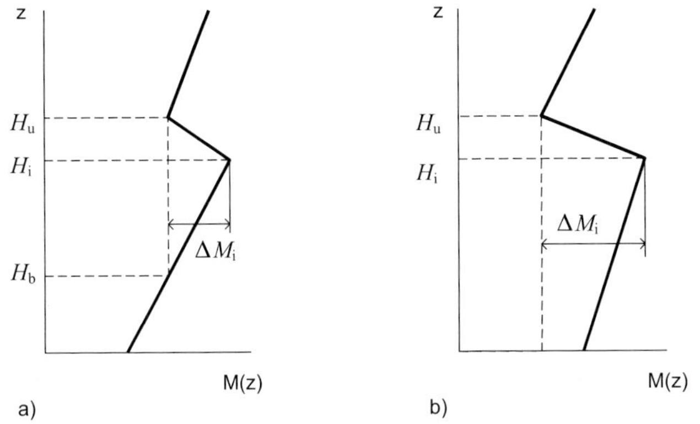

Figure 1.9 Elevated M-inversion: a) Elevated tropospheric duct in the height interval $H_u > z > H_b$. b) Surface based tropospheric duct in the height interval $H_u > z > 0$.

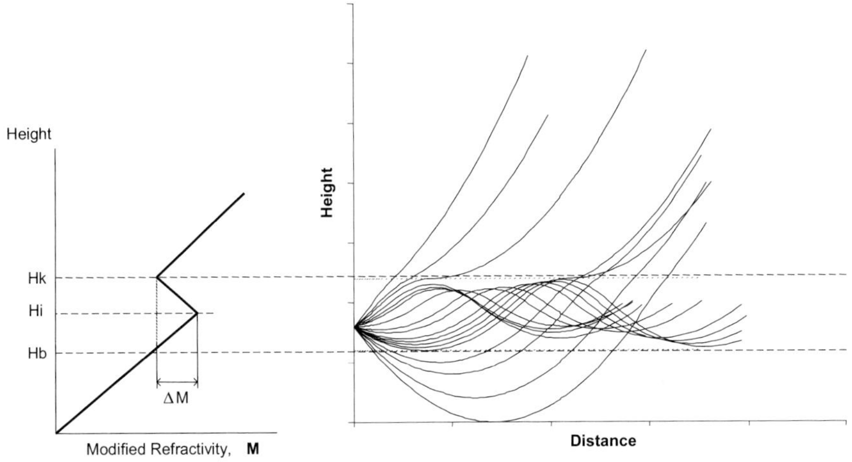

Figure 1.10 Ray traces in the elevated duct.

The observations indicate that characteristic values for the depression layers are as follows: $H_i \approx 200 - 1500$ m, $\Delta H \approx 100 - 300$ m, and $\Delta M_i \approx 10 - 30$ units. The depression layers are normally observed as stable atmospheric formations existing over a wide area of the ocean surface. The height of inversion usually coincides with the height of the atmospheric boundary layer. The elevated M-inversion normally separates a convective boundary layer from the free atmosphere above, and in the direct vicinity of M-inversion turbulence alternates with laminar flow. Due to a penetration of convective elements from the lower atmospheric boundary layer the local temperature and gradients of the wind speed increase and as a result Kelvin–Helmholtz waves develop. When these waves collapse they generate an intense turbulent structure.

Characteristic scales of the turbulent inhomogeneities in the depression layer can be estimated as follows: horizontal scale $L_{\|} \sim 500$–1000 m, vertical scale $L_z \sim 100$ m.

There is another important type of elevated M-inversion which is often formed in coastal regions due to the formation of a surface duct as a result of heating of the sea surface at sunrise and elevation of the surface M-inversion at night time. The elevated M-inversion is then unstable due to a developed convection.

1.3 Random Component of Dielectric Permittivity

A comprehensive theory of random fields and methods of their study are provided in Refs. [2] and [4]. Below we summarize the basic assumptions to be used further in this book to describe the random field of dielectric permittivity $\varepsilon(\vec{r}, t)$.

When the random field $\delta\varepsilon(\vec{r}, t)$ is stationary and uniform in time and space respectively, a two-point statistical characteristic of this field can be described via a space–time correlation function

$$B(\vec{\rho}, \tau) = \langle \delta\varepsilon(\vec{r} + \vec{\rho}, t + \tau)\delta\varepsilon(\vec{r}, t) \rangle. \tag{1.13}$$

It is assumed that $< \delta\varepsilon > = 0$, and, as a consequence of uniformity and stationarity, $\delta\varepsilon(\vec{r}, t)$ is invariant to an offset in the space and time domain, and its correlation function (1.13) depends only on the difference in space $\vec{\rho}$ and time τ between two points of observation.

The space–time correlation function $B(\vec{\rho}, \tau)$ is related to the spectrum density $G(\omega, k)$ of the field $\delta\varepsilon(\vec{r}, t)$ via a Fourier transform:

$$B(\vec{\rho}, \tau) = \int_{-\infty}^{\infty} d\omega \int d\vec{\kappa} G(\omega, \vec{\kappa}) \cdot e^{j(\vec{\kappa}\vec{\rho} - \omega\tau)}. \tag{1.14}$$

The function $G(\omega, k)$ is also called a $\omega\vec{\kappa}$ density, where $\vec{\kappa}$ is the wave vector of fluctuation in $\delta\varepsilon$. The reverse transform is given by

$$G_\varepsilon(\omega, \vec{\kappa}) = \frac{1}{(2\pi)^3} \int d^3\vec{\rho} \int_{-\infty}^{\infty} d\tau B(\vec{\rho}, \tau) e^{-j(\vec{\kappa}\vec{\rho} - \omega\tau)}. \tag{1.15}$$

Both the spatial spectrum Φ_ε and the frequency (energy) $S(\omega)$ spectrum of the fluctuations $\delta\varepsilon$ can be expressed via $\omega\vec{\kappa}$ density:

$$\Phi_\varepsilon(\vec{\kappa}) = \int_{-\infty}^{\infty} d\omega G(\omega, \vec{\kappa}) = \int_{-\infty}^{\infty} d^3\vec{\rho} B(\vec{\rho}, 0) \cdot e^{-j\vec{\kappa}\vec{\rho}}, \tag{1.16}$$

$$S_\varepsilon(\omega) = \int_{-\infty}^{\infty} d\vec{\kappa} G(\omega, \vec{\kappa}) = \int_{-\infty}^{\infty} d\tau B(0, \tau) \cdot e^{j\omega\tau}. \tag{1.17}$$

Consider the case when all time variations in $\delta\varepsilon(\vec{r}, t)$ are caused by transfer of the spatial inhomogeneties $\delta\tilde{\varepsilon}(\vec{r})$ with constant velocity $\vec{v}$, then $\delta\varepsilon(\vec{r}, t) = \delta\tilde{\varepsilon}(\vec{r} - \vec{v}t)$. If a motionless uniform random field $\delta\tilde{\varepsilon}(\vec{r})$ is described by a spatial correlation function $b_{\tilde{\varepsilon}}(\vec{\rho}) = \langle \delta\tilde{\varepsilon}(\vec{r} + \vec{\rho})\delta\tilde{\varepsilon}(\vec{r})\rangle$ then

$$B(\tau, \vec{\rho}) = b_{\tilde{\varepsilon}}(\vec{\rho} - \vec{v}\tau). \tag{1.18}$$

Substituting Eq. (1.18) into Eq. (1.15) we can find a relationship between $\omega\vec{\kappa}$ the density of the "frozen" field $\delta\varepsilon(\vec{r}, t)$ and the spatial spectrum density $\Phi_{\tilde{\varepsilon}}$ of the still field $\delta\tilde{\varepsilon}(\vec{r})$:

$$G(\omega, \vec{\kappa}) = \Phi_{\tilde{\varepsilon}}(\vec{\kappa})\delta\left(\omega - \vec{\kappa}\,\vec{v}\right). \tag{1.19}$$

Using Eqs. (1.16) and (1.19) we find that both the spatial spectrum of the frozen field $\delta\varepsilon(\vec{r}, t)$ and that of the motionless field $\delta\tilde{\varepsilon}(\vec{r})$ are the same

$$\Phi_\varepsilon(\vec{\kappa}) = \Phi_{\tilde{\varepsilon}}(\vec{\kappa}). \tag{1.20}$$

This is an important relationship since all further studies in this book are based on the hypothesis of the "frozen" turbulent fluctuations of dielectric permittivity $\delta\varepsilon(\vec{r}, t)$and all calculations are performed with a motionless field of $\delta\tilde{\varepsilon}(\vec{r})$ instead of $\delta\varepsilon(\vec{r}, t)$ and the symbol ~ can be omitted for $\delta\varepsilon(\vec{r})$.

1.3.1
Locally Uniform Fluctuations

The random field of dielectric permittivity $\delta\tilde{\varepsilon}(\vec{r})$ can be regarded as uniform only as a very rough approximation. For example, the statistical characteristics of $\delta\tilde{\varepsilon}(\vec{r})$ in the atmospheric boundary layer depend on the height above the ground and, therefore, another approach, based on the introduction of the structure function, should rather be used to describe the turbulent fluctuation of the $\delta\tilde{\varepsilon}(\vec{r})$.

The structure function is an average of the square of the module of the increment of the fluctuations $\delta\tilde{\varepsilon}(\vec{r})$ and is defined as

$$D(\vec{r}_1, \vec{r}_2) = \left\langle |\delta\varepsilon(\vec{r}_1) - \delta\varepsilon(\vec{r}_2)|^2 \right\rangle = \left\langle (\delta\varepsilon(\vec{r}_1) - \delta\varepsilon(\vec{r}_2))^2 \right\rangle. \tag{1.21}$$

The module sign is removed in Eq. (1.21), given the assumption that $\delta\tilde{\varepsilon}(\vec{r})$ has a real value. The hypothesis of the local uniformity is based on the assumption that the increment of the fluctuations of the random field $\delta\varepsilon$ at two points $\vec{r}_1$ and $\vec{r}_2$ is caused chiefly by inhomogeneities of $\delta\varepsilon$ with a size$l$ less than the distance between two points of observations: $l \leq |\vec{r}_1 - \vec{r}_2|$. While the inhomeogeneities of the large scale may exist, they produce nearly the same perturbation in the field $\delta\varepsilon$ in points $\vec{r}_1$, $\vec{r}_2$ and, therefore, a negligible difference in the perturbation in $\delta\varepsilon$. When this hypothesis holds, the field $\delta\varepsilon$ is called "locally uniform" or, using an analogy with a random process with stationary first increment, we can call $\delta\varepsilon$ a random field with uniform increment. As a consequence, the structure function $D(\vec{r}_1, \vec{r}_2)$ depends only on the spatial increment $\vec{\rho} = \vec{r}_1 - \vec{r}_2$:

$$D(\vec{r}_1, \vec{r}_2) = D(\vec{\rho}). \tag{1.22}$$

The important advantage of using a structure function is that it makes sense even when the correlation function does not exist, $B_\varepsilon(0) \to \infty$. However, if it does exist then the following relationship is valid:

$$D_\varepsilon(\vec{\rho}) = 2[B_\varepsilon(0) - B_\varepsilon(\vec{\rho})]. \tag{1.23}$$

When $B_\varepsilon(\infty) \to 0$, $D_\varepsilon(\infty) = 2B_\varepsilon(0)$ and

$$B_\varepsilon(\vec{\rho}) = \frac{1}{2}[D_\varepsilon(\infty) - D_\varepsilon(\vec{\rho})]. \tag{1.24}$$

The structure function D_ε is related to a spatial spectrum Φ_ε via the relationships

$$D_\varepsilon(\vec{\rho}) = 2\int d\vec{\kappa}\Phi_\varepsilon(\vec{\kappa})(1 - \cos\vec{\kappa}\vec{\rho}), \tag{1.25}$$

$$\Phi_\varepsilon(\vec{\kappa}) = \frac{1}{16\pi^3\kappa^2}\int_{-\infty}^{\infty} d^3\rho \cdot \vec{\kappa}\nabla D_\varepsilon(\vec{\rho})\sin(\vec{\kappa}\vec{\rho}). \tag{1.26}$$

When the random field is statistically isotropic, the increment of the fluctuations in $\delta\varepsilon$ depends only on the distance between two points $\vec{r}_1$ and $\vec{r}_2$, then $D_\varepsilon(\vec{\rho}) = D_\varepsilon(\rho)$ and $\Phi_\varepsilon(\vec{\kappa}) = \Phi_\varepsilon(\kappa)$, where $\rho = |\vec{r}_1 - \vec{r}_2|$, $\kappa = |\vec{\kappa}|$ are the modules of $\vec{\rho}$ and $\vec{\kappa}$ respectively. For locally uniform and isotropic fluctuations of $\delta\varepsilon$ Eqs. (1.25) and (1.26) can be transformed into the following ones:

$$D_\varepsilon(\rho) = 8\pi\int_0^\infty \left(1 - \frac{\sin\kappa\rho}{\kappa\rho}\right)\Phi_\varepsilon(\kappa)\kappa^2 d\kappa, \tag{1.27}$$

$$\Phi_\varepsilon(\kappa) = \frac{1}{4\pi^2\kappa^3}\int_0^\infty (\sin\kappa\rho - \kappa\rho\cdot\cos\kappa\rho)\frac{dD_\varepsilon(\rho)}{d\rho}d\rho. \tag{1.28}$$

The Eqs. (1.19) and (1.20) are valid for a "frozen" and locally uniform random field $\delta\varepsilon(\vec{r}, t)$. The equation for the time–space correlation function (1.18) can be substituted by a similar equation for a space–time structure function

$$D_\varepsilon(\vec{\rho}, \tau) = D_\varepsilon(\vec{\rho} - \vec{v}\tau). \tag{1.29}$$

The turbulent fluctuations of dielectric permittivity $\delta\varepsilon(\vec{r})$ in a free atmosphere can be modelled as a locally uniform random field. The structure function $D_\varepsilon(\rho)$ of fluctuations $\delta\varepsilon$ obeys a law of 2/3 in the Kolmogorov–Obukhov model [4] for large enough ρ:

$$D_\varepsilon(\rho) \approx C_\varepsilon^2 \rho^{2/3}, \quad \rho >> l_0 \tag{1.30}$$

where C_ε^2 is the structure constant, a measure of fluctuation intensity and l_0 is the internal scale of the turbulence (the inhomeogeneties with the scale less than l_0 dissipate rapidly). For small ρ the structure function follows the square law:

$$D_\varepsilon(\rho) \approx C_\varepsilon^2 \rho^2, \quad \rho << l_0. \tag{1.31}$$

As observed from Eq. (1.30) the structure function $D_\varepsilon(\rho)$ is unlimited by a scale of fluctuations and allows the existence of turbulent bursts of any scale, which is contradictory to the applicability of the Kolmogorov–Obukhov model which is itself limited to an inertial interval of the locally isotropic turbulence.

In order to limit the external scales of the fluctuations to an inertial interval of the turbulence and to make the model consistent one can introduce a maximum scale L_0 of the turbulent fluctuations and therefore a "saturation" feature in the behavior of the structure function:

$$D_\varepsilon(\rho) = C_\varepsilon^2 L_0^{2/3}, \quad \rho >> L_0\,. \tag{1.32}$$

In fact, a saturation means a limited value of $D_\varepsilon(\infty)$ and, therefore, a limited value of variance $\sigma_\varepsilon^2 = B_\varepsilon(0)$ of the fluctuations $\delta\varepsilon$. In this case those values are related as follows: $C_\varepsilon^2 L_0^{2/3} = 2\sigma_\varepsilon^2$.

A combined structure function for locally uniform and isotropic fluctuation in $\delta\varepsilon$can be written as follows [4]:

$$D_\varepsilon(\rho) = \begin{cases} C_\varepsilon^2 l_0^{-4/3} \rho^2, & \rho \ll l_0 \\ C_\varepsilon^2 \rho^{2/3}, & l_0 \ll \rho \ll L_0 \\ C_\varepsilon^2 L_0^{2/3}, & \rho >> L_0 \end{cases} \tag{1.33}$$

Within an inertial interval of locally isotropic turbulence the spatial spectrum $\Phi_\varepsilon(\vec{\kappa})$ is described by the Kolmogorov–Obukhov model

$$\Phi_\varepsilon(\vec{\kappa}) = 0.033 C_\varepsilon^2 \kappa^{-11/3} \exp\left(-\frac{\kappa^2}{\kappa_m^2}\right). \tag{1.34}$$

Here $\vec{\kappa}$ is the wave vector of fluctuation in $\delta\varepsilon$, $\kappa = |\vec{\kappa}|$, C_ε^2 is the structure constant, a measure of the fluctuation intensity, $\kappa_m = 5.92/l_0$ and l_0 is the internal scale of the turbulence.

The applicability of Kolmogorov's model (1.34) for the atmospheric boundary layer over the ocean surface has been studied in Ref. [5]. The basic observations are as follows:

- A one-dimensional spectrum of the turbulent fluctuations in the horizontal plane follows Kolmogorov's model up to scales: $L_{\|}$~10–100 m. The spectrum interval over which the postulates of a locally isotropic turbulence are valid depends on the stability of the atmosphere and decreases greatly when the stability intensifies. In the case of very stable stratification associated with the presence of M-inversions the turbulent structure reveals significant anisotropy of the large-scale fluctuations.
- At heights above the surface comparable with the heights of the swells, the state of the sea surface reveals a significant impact on a form of the spectrum of turbulent fluctuations. This manifest itself in the form of flares in spectral density at the frequencies corresponding to the energy-bearing frequencies of the sea waves. The amplitude of the wave perturbations decreases with height and has no practical effect on the form of the spectrum at heights of 15–20 m above the surface;
- The intensity of the high-frequency turbulent fluctuations depends parametrically on the height above the surface. In a low-frequency domain (the buoyancy and energy intervals) one must include additional parameters in the model of spectrum: the thickness of the atmospheric boundary layer and some parameters characterizing the roughness of the sea surface.

It can be stated that despite numerous studies, a physically justifiable model for a three-dimensional spectrum of fluctuations in the index of refraction within the atmospheric boundary layer does not yet exist. The anisotropy of the spatial fluctuations in the refractive index is commonly introduced via the anisotropy parameter $\alpha = L_z/L_{\|}$ in the following model of the spectrum:

$$\Phi_\varepsilon(\vec{\kappa}) = \frac{0.063\,\sigma_\varepsilon^2 L_z L_{\|}^2}{\left(1 + \kappa_{\|}^2 L_{\|}^2 + \kappa_z^2 L_z^2\right)^{11/6}} \tag{1.35}$$

where the σ_ε^2 is a variance of the fluctuations. It can be noted that in the case of turbulence $L_z, L_{\|} \to \infty$, $\sigma_\varepsilon^2 \to \infty$. The generalisation of Kolmogorov's model to anisotropic turbulence can be given as follows:

$$\Phi_\varepsilon(\vec{\kappa}) = \frac{0.033\,C_{\varepsilon\perp}^2\,\alpha}{\left(\kappa_\perp^2 + \alpha^2\kappa_z^2\right)^{11/6}} = \frac{0.033\,C_{\varepsilon z}^2\,\alpha^{5/3}}{\left(\kappa_\perp^2 + \alpha^2\kappa_z^2\right)^{11/6}} \tag{1.36}$$

where $C_{\varepsilon\perp}$ and $C_{\varepsilon z}$ are structure constants in the horizontal plane and over height respectively, as observed they are not independent and are related via

$$C_{\varepsilon\perp} = \alpha^{1/3} C_{\varepsilon z}. \tag{1.37}$$

In practical measurements the time of averaging is always finite and so variance of the fluctuations σ_ε^2 and scales $L_\perp$, L_z is observed. In fact model (1.35) can be trans formed into model (1.36) with the substitution $\sigma_\varepsilon^2 = 1.9 C_{\varepsilon\perp}^2 L_\perp^{2/3}$, assuming that the external scale of the fluctuations in dielectric permittivity $L_\perp$ is confined within the micro-range region of the spectrum of turbulence.

The observations [6] suggest that the value of the anisotropy parameter α itself is a function of the external scale of the turbulent fluctuations and can be estimated to be of the order of $\alpha \sim 0.1 - 0.2$ for $L_z \sim 1$ m at the heights close to the water level. At the upper boundary of ABL the parameter α may vary from $\alpha \sim 0.8$ for $L_\parallel \sim 15$ m to $\alpha \sim 0.005$ for $L_\parallel \sim 10{,}000$ m.

References

1 Jensen, N.O., Lenshow, D.H. An observational investigation on penetrative convection. *J.Atmos.Sci.*, 1978, 35(10), 1924–1933.

2 Monin, A.S., Yaglom, A.M. *Statistical Fluid Mechanics*, Vol.1, MIT Press, Cambridge, MA, 1971, p. 769.

3 Gossard, E.E. Clear weather meteorological effects on propagations at frequencies above 1 GHz, *Radio Sci.*, 1981,16(5), 589–608.

4 Tatarskii, V.I. *The Effects of Turbulent Atmosphere on Wave Propagation*, IPST, Jerusalem, 1971.

5 Gavrilov, A.S., Ponomareva, S.M. *Turbulence Structure in the Ground Level Layer of the Atmosphere.* Collected Data, Meteorology Series, No.1, Research Institute for Meteorological Information, Obninsk (in Russian), 1984.

6 Kukushkin, A.V., Freilikher, V.D. and Fuks, I.M. Over-the-horizon propagation of UHF radio waves above the sea , *Radiophys. Quantum Electron.* (translated from Russian), Consultant Bureau, New York , RPQEAC 30 (7), 1987, 597–620.

2
Parabolic Approximation to the Wave Equation

2.1
Analytical Methods in the Problems of Wave Propagation in a Stratified and Random Medium

The analogy between the non-stationary problem of quantum mechanics and the parabolic approximation in a problem wave propagation was likely first explored by Fock [1]. It is well known that the methods of classic wave theory were utilised in quantum mechanics in the early stages of its development. At present, the situation is rather the reverse, and the mathematical methods of quantum mechanics have become widely adopted in the latest developments of wave theory.

The asymptotic solution to the problem of the diffraction of a plane monochromatic wave over a sphere of large radius, compared with the wavelength of radiation, was obtained by Fock in 1945 [2]. In obtaining this solution, he developed an asymptotic theory of diffraction on the basis of Leontovich's boundary conditions [3] and the concept of the "local" field [2]. The approach utilised several large parameters involved in problem formulation, such as $|\eta| >> 1$, $m = (ka/2)^{1/3} >> 1$, $\eta = \varepsilon_g + j4\pi\sigma/\omega$, ε_g, σ, the dielectric permittivity and conductivity of the earth, respectively, ω is the cyclic frequency of the radiated wave, $k = 2p/\lambda$ is a wavenumber, λ is a wavelength, a is the radius of the earth's sphere. The same results were obtained by Fock later in Ref. [4] by means of the parabolic approximation to the wave equation which is based on significantly different parameters, with $m >> 1$, and the characteristic scales of the wave oscillations in two different directions: along the earth's surface and normal to it. With regards to radio wave propagation at a frequency above 1 GHz the important development was a solution of the parabolic equation for the stratified troposphere, where the refractive index depends solely on the height above the ground surface [5]. The analysis of the wave field at low altitudes z, which are small compared with the radius of the earth's curvature a, is convenient to perform by means of the introduction of the modified refractivity: $n_m(z) = n(z) + z/a$. The solution obtained in Ref. [5] is suitable for the very general case of the dependence n(z) and impedance boundary conditions at the earth's surface. The solution in Ref. [5] is represented by a contour integral which in the shadow region is calculated by a sum over residues in the poles of the integrand, i.e. a series of normal waves [6]. In the line-of-sight (LOS) region the contour integral is

Radio Wave. Alexander Kukushkin

ISBN: 3-527-40458-9

commonly calculated using stationary phase methods similar to a geometric optic approximation [1, 7]. Further study of the wave propagation in a stratified troposphere was dedicated to the development of the effective methods of the field representation for various height profiles of the refractivity and combinations of the relative location of the transmitting and receiving antennae [8–16]. An alternative analytical method to the solution of wave propagation in stratified medium, the method of "invariant submergence", was developed by Klatskin [17]. Using that method, a boundary problem, which is, in fact, a problem of diffraction, can be transformed into a problem of evolution. A significant advantage of this method is its inherent capability to study the wave propagation through random media, and the method is especially effective when applied to a randomly stratified medium.

Together with a ducting mechanism, the wave scattering on random inhomogeneities of the refractive index plays a significant role in the study of radio wave propagation in the atmospheric boundary layer at frequencies above 1 GHz. The straightforward application of either the contour integral or normal mode series to a problem of wave propagation in a random troposphere will face significant difficulties. The major problem will be related to the divergence of the matrix elements of the normal wave transformation due to a scattering on random fluctuations in the refractive index. The deployment of the method of "invariant submergence" also does not lead to an analytical solution to the problem though numerical calculation is possible.

In this book we use the representation of the Green function in the form of an expansion over the eigen functions of a continuous spectrum which allows one to avoid the above mentioned difficulties of divergence of matrix elements and then to separate the matrix coefficients with a "nearly discrete" and continuum spectrum with normalised height gain functions. This approach led to some advances in solving several problems, in particular, the problem of wave propagation in a tropospheric duct filled with random irregularities of the refractive index in addition to a regular supper-refractive gradient of the refractivity. The disadvantage of the method is that the separation of the discrete spectrum of eigen functions is performed in the "unperturbed", i.e. regular, part of the Schrödinger operator and, therefore, the spectrum of eigen functions is left unchanged by the random component of the potential term. The study of the spectrum of Schrödinger's equation with random potential is a complex problem by itself and most advances so far have been achieved in the one-dimensional problem [18].

The theory of wave scattering in a random medium has been developed during the last decades [19–23]. The most significant progress has been achieved in the solution to the problems of wave propagation in either a random but unbound medium or a uniform medium with a random boundary interface. In the solution of these problems a statistical approach is commonly used. This approach is targeted at obtaining the statistical characteristics of the scattered field: probability function or statistical moments of the field. Among them the first two moments: average field (coherent component) and either coherence function or field intensity, are of significant interest in many practical applications. There are three known methods of obtaining the closed equations for the field moments: Feynman diagrams [19, 24, 25], local perturbation [26–28] and the Markov process approximation [29–31].

The study of a coherent component can be done using the modified perturbation theory [32, 33], a similar result is obtained from the diagram methods [25]. The advances in the solution of the higher order moments of the field have been obtained after application of the parabolic approximation to the wave equation [21, 22]. Neglecting back scattering in the parabolic approximation allows one to employ the concept of the Markov process [29–31] and all the following analogies from the statistical theory of Markov processes.

The path integral in the problem of wave propagation in a random medium was introduced in Refs. [34–36]. The advantage of the functional approach, as noted in Ref. [22], is that the formal solution to the problem is written via quadrature. The moment of the field of any order can, in principle, be obtained by averaging the respective dynamic solution which is described by a multiple functional integral in the space of the virtual trajectories. The solutions to the first two moments of the field obtained from both the method of closed equations for the moments and the method of the path integral provide exactly the same result, as shown in Refs. [22, 36] under conditions of the Markov process approximation. The study of the field moments of order higher than two is easier with the use of the path integral, since this approach does not formally require a solution of the equations for the field moments. In fact, this advantage became apparent in the calculation of the moments of the intensity (signal strength), which are not described by a closed equation.

The current limitations of the path integral method could be rather associated with the still limited mathematical methods of analytical calculation of the trajectory integrals, despite the serious studies in that area [37, 38].

Finally, we comment on studies concerning scattering on a rough sea surface. The basic conclusion is that, until now, there is no adequate theory that takes into account the combined effect of refraction, diffraction and scattering on random inhomegeneities of the refractive index in the volume of the tropospheric layer as well as scattering on a rough sea surface. The impact from a rough surface cannot be regarded as negligible in the general case, at least from a theoretical point of view. Known attempts are limited to the deterministic model of the sea surface or a semi-empirical approach in treating the scattering on the sea surface in the Kirchhoff approximation [39, 40] that is applicable to the coherent component of the scattered field. While the theory of wave scattering on a random surface is an established science by itself, see for instance Refs. [19, 41], the studies known to the author were concerned with scattering theory on a random surface while treating the propagation media in a very simplistic way, basically, as deterministic and uniform. In applying the scattering theory to a ducting mechanism, the approach developed in Refs. [19, 41] should eventually be modified in order to take into account the multiple effects of refraction, diffraction and scattering. As discussed in Chapter 1, the sea waves could be responsible for the modulation of the turbulence spectrum in the near-surface layer, where the major action of radio propagation takes place. On the other hand, all previous advances in theory have been driven by some unresolved problems in experiments on the radio wave propagation phenomenon. So far, the currently available radio coverage prediction system [39] is reported to be in reasonably good agreement with the results of existing models, at least for measured signal strength.

2.2 Parabolic Approximation to a Wave Equation in a Stratified Troposphere Filled with Turbulent Fluctuations of the Refractive Index

Let us consider the vertical electric dipole in a spherical coordinate system (r, ϑ, ϕ) with the spherical axis coming through the dipole. Assume that the dielectric permeability ε is a function of the radius only, e.g. $\varepsilon = \varepsilon(r)$. As shown in Ref. [1], the classical solution to the electromagnetic field can be derived by introduction of Debye's potentials U and V. The field components are given by the following equations:

$$\begin{aligned} E_r &= \frac{1}{r}\Delta^* U, \\ E_\vartheta &= -\frac{1}{\varepsilon r}\frac{\partial^2}{\partial r \partial \vartheta}(\varepsilon r U) + j\frac{k}{\sin\vartheta}\frac{\partial V}{\partial \phi}, \\ E_\phi &= -\frac{1}{\varepsilon r \sin\vartheta}\frac{\partial^2}{\partial r \partial \phi}(\varepsilon r U) - jk\frac{\partial V}{\partial \vartheta}, \end{aligned} \tag{2.4}$$

$$\begin{aligned} H_r &= -\frac{1}{r}\Delta^* V, \\ H_\vartheta &= j\frac{k\varepsilon}{\sin\vartheta}\frac{\partial U}{\partial \phi} + \frac{1}{r}\frac{\partial^2 (rV)}{\partial r \partial \vartheta}, \\ H_\phi &= -jk\varepsilon\frac{\partial U}{\partial \vartheta} + \frac{1}{r\sin\vartheta}\frac{\partial^2 (rV)}{\partial r \partial \phi}. \end{aligned} \tag{2.5}$$

We assume the field to be a harmonic function of time – $\exp(-j\omega t)$, ω is a cyclic frequency of the electromagnetic radiation, $k = \omega/c$ is a wave number, and c is the speed of light. The notation Δ^* represents a Laplace operator on a sphere

$$\Delta^* = \frac{1}{\sin\vartheta}\frac{\partial}{\partial \vartheta}\left(\sin\vartheta\frac{\partial}{\partial \vartheta}\right) + \frac{1}{\sin^2\vartheta}\frac{\partial^2}{\partial \phi^2}. \tag{2.6}$$

The Maxwell equations

$$\begin{aligned} \mathrm{rot}\vec{H} &= jk\varepsilon\vec{E}, \\ \mathrm{rot}\vec{E} &= -jk\vec{H} \end{aligned} \tag{2.7}$$

will be satisfied if potentials U and V obey the following equations:

$$\frac{1}{r}\frac{\partial}{\partial r}\left(\frac{1}{\varepsilon}\frac{\partial(\varepsilon r U)}{\partial r}\right) + \frac{1}{r^2}\Delta^* U + k^2\varepsilon U = 0, \tag{2.8}$$

$$\frac{1}{r}\frac{\partial^2 (rV)}{\partial r^2} + \frac{1}{r^2}\Delta^* V + k^2\varepsilon V = 0. \tag{2.9}$$

The approximate Leontovitch's boundary conditions for U and V have the form

$$\frac{\partial(\varepsilon rU)}{\partial r} = -jk\frac{\varepsilon}{\sqrt{\eta}}rU, \tag{2.10}$$

$$\frac{\partial(rV)}{\partial r} = -jk\sqrt{\eta}rV \tag{2.11}$$

with $r = a$, where a is the earth's radius.

Let us assume that dielectric permittivity $\varepsilon(\vec{r})$ is stratified on average, i.e. $\langle\varepsilon(\vec{r})\rangle = \varepsilon_0(r)$. This means that $\varepsilon(\vec{r})$ is a random function of the space coordinates $\vec{r} = \{r, \vartheta, \phi\}$: $\varepsilon(\vec{r}) = \varepsilon_0(r) + \delta\varepsilon(\vec{r})$, where $\delta\varepsilon(\vec{r})$ is a random component present in each realisation of $\varepsilon(\vec{r})$, and $\langle\delta\varepsilon(\vec{r})\rangle = 0$. Assume also that $|\delta\varepsilon/\varepsilon| << 1$. The angular brackets here and below mean averaging over the statistical ensemble of the fluctuations $\delta\varepsilon$.

Let us assume that, the relationships between field components $\vec{E}$, $\vec{H}$ and potentials U, V are still given by Eqs. (2.4) and (2.5). Strictly speaking, the Maxwell equations (2.7) cannot be satisfied for an arbitrary function $\delta\varepsilon(\vec{r})$, and one can seek approximation to Eq. (2.7) where the divergent terms are small.

Introduce vector $\vec{x} = \{a\vartheta, \phi r\sin\vartheta\}$. The divergent terms in Eq. (2.7) will be small when

$$\left|\frac{\partial\varepsilon}{\partial x_i}\right| << k \tag{2.12}$$

where x_i is any of component of the vector $\vec{x}$. When the inequality (2.12) holds, the potentials U and V are governed by Eqs. (1.8), (1.9)where $\varepsilon = \varepsilon_o(r) + \delta\varepsilon(\vec{r})$.

Let us examine the vertically polarised field for which $V = 0$, $U \neq 0$. For high frequencies when $ka >> 1$, we can select a "dedicated" direction of wave propagation $x = a\vartheta$ aligned with the direction along the arc of the earth's radius. With $ka >> 1$, the radial component of the electric field E_r is related to the potential U via: $E_r \approx -ka^2 U$.

Following Ref. [1], introduce a new function

$$U_1 = \varepsilon rU \tag{2.13}$$

which is governed by the equation

$$\frac{\partial}{\partial r}\left(\frac{1}{\varepsilon}\frac{\partial U_1}{\partial r}\right) + \frac{1}{\varepsilon r^2}\left[\begin{matrix}\dfrac{\partial^2 U_1}{\partial\vartheta^2} - \dfrac{\partial U_1}{\partial\vartheta}\left(\dfrac{2}{\varepsilon}\dfrac{\partial\varepsilon}{\partial\vartheta} - \dfrac{\cos\vartheta}{\sin\vartheta}\right) + \\ U_1\left(\dfrac{1}{\varepsilon^2}\left(\dfrac{\partial\varepsilon}{\partial\vartheta}\right)^2 - \dfrac{1}{\varepsilon}\dfrac{\partial^2\varepsilon}{\partial\vartheta^2} - \dfrac{\cos\vartheta}{\varepsilon\sin\vartheta}\dfrac{\partial\varepsilon}{\partial\vartheta}\right)\end{matrix}\right]$$

$$+\frac{1}{\varepsilon r^2\sin\vartheta}\left[\frac{\partial^2 U_1}{\partial\varphi^2} - \frac{2}{\varepsilon}\frac{\partial U_1}{\partial\varphi}\frac{\partial\varepsilon}{\partial\phi} - \frac{U_1}{\varepsilon^2}\left(\frac{\partial^2\varepsilon}{\partial\varphi^2} - \frac{2}{\varepsilon}\left(\frac{\partial\varepsilon}{\partial\varphi}\right)^2\right)\right] + k^2 U_1 = 0 \tag{2.14}$$

and obeys the boundary condition

$$\frac{\partial(U_1)}{\partial r} = -jk\frac{U_1}{\sqrt{\eta+1}}, \quad \text{with } r = a. \tag{2.15}$$

Let us isolate the slow varying complex amplitude W_1 in the wave field by

$$U_1 = W_1 e^{jka\vartheta} \tag{2.16}$$

and introduce coordinate $y = (r\sin\vartheta)\varphi$. The amplitude W_1 is then given by equation

$$\begin{aligned}
&\frac{\partial^2 W_1}{\partial r^2} + \frac{2jk}{r}\frac{\partial W_1}{\partial\vartheta} + \frac{\partial^2 W_1}{\partial y^2} + \left(k^2\varepsilon - k^2\frac{a^2}{r^2}\right)W_1 = \\
&\frac{\partial\varepsilon}{\partial r}\frac{\partial W_1}{\partial r} - \frac{1}{r^2}\frac{\partial^2 W_1}{\partial\vartheta^2} + j\frac{k}{r}W_1\left(\frac{2}{\varepsilon}\frac{\partial\varepsilon}{\partial\vartheta} - \frac{\cos\vartheta}{\sin\vartheta}\right) + \\
&\frac{W_1}{r^2}\left(\frac{1}{\varepsilon^2}\left(\frac{\partial\varepsilon}{\partial\vartheta}\right)^2 - \frac{1}{\varepsilon}\frac{\partial^2\varepsilon}{\partial\vartheta^2} - \frac{\cos\vartheta}{\varepsilon\sin\vartheta}\frac{\partial\varepsilon}{\partial\vartheta}\right) + \\
&\frac{1}{r^2}\left[\frac{2}{\varepsilon}\frac{\partial W_1}{\partial y}\frac{\partial\varepsilon}{\partial y} - \frac{W_1}{\varepsilon}\left(\frac{\partial^2\varepsilon}{\partial y^2} - \frac{2}{\varepsilon}\left(\frac{\partial\varepsilon}{\partial y}\right)^2\right)\right]
\end{aligned} \tag{2.17}$$

at the right-hand side of which we have a correction term. Using qualitative arguments, it can be shown that in the case of a stratified medium $\varepsilon = \varepsilon(r)$ the terms on the left have an order of magnitude of $k^2 W_1/m^2$ while on the right the terms containing derivatives are of the order of $k^2 W_1/m^4$. The right-hand side terms containing $\sin\vartheta$ in the denominator will also be small if the following inequality holds

$$x >> \frac{a}{m}. \tag{2.18}$$

The problem now is to evaluate the terms in the right-hand side of Eq. (2.17) for the case of random $\delta\varepsilon(\vec{r})$ caused by turbulence. Let us introduce coordinate $z = r - a$. The characteristic scales L of variations in W_1 over x, y, z can be estimated as follows:

$$L_x \sim \frac{a}{m}, \quad L_y \sim \sqrt{\lambda x}, \quad L_z \sim \frac{m}{k}. \tag{2.19}$$

respectively. Having compared the terms in the left- and right-hand sides of Eq. (2.17) we obtain the inequalities

$$\left(\frac{\partial\varepsilon}{\partial x}\right)^2 << \frac{m^2}{a^2}, \quad \left(\frac{\partial\varepsilon}{\partial y}\right)^2 << \frac{1}{\lambda x}, \quad \left(\frac{\partial\varepsilon}{\partial z}\right)^2 << \frac{k^2}{m^2} \tag{2.20}$$

which, being satisfied, allow one to treat the terms in the right-hand side of Eq. (2.17) as negligible. Therefore, with fulfilment of inequalities (2.20) the right-hand side of Eq. (2.17) can be replaced with zero.

The inequalities (2.20) put limitations chiefly on the intensity of small-scale fluctuations $\delta\varepsilon(\vec{r})$. We can assume that the fluctuations in dielectric permittivity $\delta\varepsilon(\vec{r})$

are locally uniform and isotropic. Let us introduce a spatial spectrum Φ_ε of fluctuations $\delta\varepsilon(\vec{r})$ by the equation

$$\Phi_\varepsilon(\vec{\kappa}) = \frac{1}{16\pi^3} \int d^3\vec{\rho} \cdot \text{grad}\left\langle [\delta\varepsilon(\vec{r}+\vec{\rho}) - \delta\varepsilon(\vec{r})]^2 \right\rangle \frac{\vec{\kappa}\sin(\vec{\kappa}\vec{\rho})}{\kappa} \tag{2.21}$$

where $\vec{\kappa} = \{\kappa_x, \kappa_y, \kappa_z\}$ is the wave vector of the fluctuations in the dielectric permittivity, $\kappa = |\vec{\kappa}|$.

We can evaluate the gradients in $\delta\varepsilon(\vec{r})$ by

$$\left\langle \left(\frac{\partial\varepsilon}{\partial x_n}\right)^2 \right\rangle \sim \int d^3\vec{\kappa} \cdot \kappa_n^2 \Phi_\varepsilon(\vec{\kappa}) \tag{2.22}$$

where x_n is any of the coordinates x, y, z. For the spatial spectrum of fluctuations $\delta\varepsilon(\vec{r})$ we can use the formula related to the equilibrium interval of the locally-isotropic turbulence [42]:

$$\Phi_\varepsilon(\vec{\kappa}) = 0.033 C_\varepsilon^2 \kappa^{-11/3} \exp\left(-\frac{\kappa^2}{\kappa_m^2}\right) \tag{2.1}$$

where c_z is a structure constant, $\kappa_m = 5.92/l_0$, l_0 is an internal scale of turbulence. Substituting Eq. (2.1) into Eq. (2.20) we obtain

$$0.033\, C_\varepsilon^2 \kappa_m^{4/3} << \frac{k}{x}, \frac{k^2}{m^2}, \frac{m^2}{a^2}. \tag{2.24}$$

Assuming reasonably short distances and high frequencies of radiation, e.g. $m >> 1$ and $x \le 10^3$ km, we can see that the major impact comes from variations of the z-coordinate, and the inequality (2.24) thus reduces to

$$0.033\, C_\varepsilon^2 \pi \left(\frac{a}{2}\right)^{2/3} \left(\frac{\lambda}{l_0}\right)^{4/3} << 1. \tag{2.25}$$

Here we took into account that $k/m = k^{2/3}(a/2)^{1/3}$. Estimating the parameters $C_\varepsilon^2 = 10^{-14}\text{cm}^{-2/3}$, $a = 8500$ km, we observe that the coefficient with $(\lambda/l_0)^{4/3}$ has an order of 10^{-9}.

Now we can conclude that with fulfilment of inequalities (2.18), (2.25) and $z/a << 1$, the complex amplitude $W_1(\vec{r})$ obeys the approximate equation

$$j2k\frac{\partial W_1}{\partial x} + \frac{\partial^2 W_1}{\partial y^2} - \frac{\partial^2 W_1}{\partial z^2} + k^2(\varepsilon_m(z) - 1 + \delta\varepsilon(\vec{r}))W_1 = 0 \tag{2.26}$$

and boundary condition

$$\frac{\partial W_1}{\partial z} = -jk\frac{W_1}{\sqrt{\eta+1}} \quad \text{with } z = 0. \tag{2.27}$$

The potential U is given by

$$U(\vec{r}) = \frac{e^{jka\vartheta}}{\varepsilon r} W_1(\vec{r}). \tag{2.28}$$

The next step is to figure out the correct representation of the field amplitude W_1 at small distances from the source in order to obtain an initial condition. In particular, at the distance where the curvature of the earth can be neglected as well as the ray's refraction we have to obtain a reflection formula in the parabolic approximation [1]. Introducing the coordinates of the source: $x = 0$, $y = y_0$, and $z = z_0$, and assuming that $\eta >> 1$ the reflection formula will be satisfied when

$$W_1(x, \vec{\rho}) = 2\frac{\varepsilon(x, \vec{\rho})}{x} a \exp\left\{j\frac{k}{x}(\vec{\rho} - \vec{\rho}_0)^2\right\} \tag{2.29}$$

where $\vec{\rho} = \{y, z\}$, $\vec{\rho}_0 = \{y_0, z_0\}$.

Taking into account the relationship

$$\lim|_{x\to x'} \frac{k}{2\pi(x-x')} \exp\left\{j\frac{k}{2(x-x')}\left(\vec{\rho} - \vec{\rho}'\right)^2\right\} = \delta\left(\vec{\rho} - \vec{\rho}'\right)$$

where $\delta(\vec{\rho})$ is a Dirac's delta-function defined by

$$\delta(\vec{\rho}) = \frac{1}{4\pi^2}\int d^2\vec{\kappa} \exp(j\vec{\kappa}\vec{\rho}), \tag{2.30}$$

we make a transition in Eq. (2.29) with $x \to 0$ and obtain an expression for singularity at the source

$$\lim|_{x\to 0} W_1(x, \vec{\rho}) = 4\pi j\varepsilon(0, \vec{\rho}_0)\frac{a}{k}\delta(\vec{\rho} - \vec{\rho}_0). \tag{2.31}$$

Equation (2.29) shall take place in a line-of-sight region, i.e. with $x << \sqrt{2az_0}$. Taking into account that the inequality (2.18) is to hold as well, we obtain the limitations at the height of the source z_0 under which the expression (2.31) is valid:

$$z_0 >> a/m^2\,. \tag{2.32}$$

It should be noted that in the case of very small heights of the source, another approach can be used and the solution to Eq. (2.26) should be tailored with the solution of Veil-Van-der-Paul [1] at small distances x.

In conclusion, one should note that there is some ambiguity with the definition of the complex amplitude via Debye's potential U. Indeed, equations similar to Eq. (2.26) can be obtained for the other functions W_2 and W_3 defined by the following expressions:

$$\begin{aligned} W_2 &= \varepsilon r\sqrt{\sin\vartheta}\cdot e^{-jka\vartheta}U, \\ W_3 &= \varepsilon r\sin\vartheta e^{-jka\vartheta}U. \end{aligned} \tag{2.33}$$

The area of applicability of the parabolic approximation for W_3 is defined by the same set of inequalities (2.18), (2.24) while for W_3 the inequality (2.18) can be replaced by a less stringent one:

$$x >> m/k. \tag{2.34}$$

Let us summarize the major steps undertaken in this section:

1. The components of the electromagnetic field $\vec{E}$ and $\vec{H}$ can be represented via Debye's potentials U and V. Even in the presence of a random turbulent component of the refractive index the de-polarization effects can be neglected and, therefore, the field components are governed by independent equations for U and V.
2. The wave equations for both potentials U and V can be approximated by parabolic equations for a slow varying complex amplitude. In this approximation all radiated waves are travelling only in one direction, away from the source. The Leontovich boundary conditions are used along with the parabolic equation. The conditions in the source are also modified in a parabolic approximation.
3. During the transformation from the Helmoltz equation to a parabolic equation we actually replaced the earth sphere with a cylinder of the same radius and then flattened the cylinder by means of compensating the earth's curvature by "ray's curvature" in the opposite direction. This is a well-known "flat earth" approximation with modified refractivity and locally Cartesian coordinates.

2.3 Green Function for a Parabolic Equation in a Stratified Medium

Let $\delta\varepsilon(\vec{r}) = 0$ in Eq. (2.26) and take into account that for the frequencies above 1 GHz the boundary condition (2.27) is reduced to $W(\vec{r}) = 0$ at $z = 0$. By definition, a Green function $G(\vec{r}, \vec{r}\,')$ is governed by the equation

$$2jk\frac{\partial G}{\partial x} + \Delta_{\perp} G + k^2 [\varepsilon_m(z) - 1]G = -4\pi\delta(\vec{r} - \vec{r}\,'), \tag{2.35}$$

where $\Delta_{\perp} = \dfrac{\partial^2}{\partial y^2} + \dfrac{\partial^2}{\partial x^2}$,

obeys the boundary condition

$$G(\vec{r}, \vec{r}\,') = 0 \quad \text{with } z = 0,$$

and is a continuous and limited function for all $\vec{r} \neq \vec{r}\,'$.
Introducing a Fourier transform over the transverse coordinates x, y:

$$G(\vec{r}, \vec{r}\,') = \frac{1}{4\pi^2} \int_{-\infty}^{\infty} dp \int_{-\infty}^{\infty} dq \widehat{G}\left(p, q, z, z'\right) \exp\left(jp(x - x') + jq(y - y')\right)$$

and taking into account the definition of the δ-function, we obtain

$$\frac{d^2 \widehat{G}}{dz^2} + k^2 \left[(\varepsilon_m(z) - 1) - 2kp - q^2\right] \widehat{G} = -4\pi\delta(z - z'). \tag{2.36}$$

We will seek a solution to Eq. (2.36) as a composition over functions $\Psi_E(z)$ which are governed by the equation

$$\frac{d^2\Psi}{dz^2} + k^2[(\varepsilon_m(z) - 1)]\Psi = E\Psi \tag{2.37}$$

and satisfy the boundary conditions

$$\Psi_E(z) = 0, \quad z = 0; \quad \Psi_E(z) = 0, \quad z \to \infty. \tag{2.38}$$

Equation (2.37) is similar to the one-dimensional Schrödinger equation for a particle in a field with potential energy which is proportional to the term $(\varepsilon_m(z) - 1)$. As known from Refs. [43, 44], in the case of unlimited potential (in Eq. (2.37) $(\varepsilon_m(z) - 1) \to 2z/a$ with $z \to \infty$)), the motion of the particle in a stationary state is infinite to $z \to +\infty$ and the spectrum of the eigen values E is purely continuous.

Therefore, the function $\widehat{G}(p, q, z, z')$ can be thought of as a composition of the eigen functions of the continuous spectrum which obey Eq. (2.37) and the boundary conditions (2.38):

$$\widehat{G}(p, q, z, z') = \int_{-\infty}^{\infty} dEB(p, q, E, z')\Psi_E(z). \tag{2.39}$$

The eigen functions of a continuous spectrum satisfy the orthogonality conditions

$$\int_{0}^{\infty} \Psi_E(z)\Psi^*_{E_1}(z)dE = \delta(E - E_1) \tag{2.40}$$

and “completeness”

$$\int_{-\infty}^{\infty} \Psi_E(z)\Psi^*_E(z') = \delta(z - z'). \tag{2.41}$$

The conditions (2.40) and (2.41) are known from quantum mechanics [44] where they are proven for potentials limited at infinite z. Using methods similar to those in Refs. [43, 44] one can prove Eqs. (2.40) and (2.41) for a potential with a linear increment at infinity. The sign $*$ means a complex conjugate, here and below.

Substituting Eq. (2.39) into Eq. (2.36) and taking into account Eqs. (2.40) and (2.41) we obtain

$$B(p, q, E, z') = -\frac{4\pi\Psi^*_E(z')}{E-2kp-q^2}, \tag{2.42}$$

and then performing the integration in Eq. (2.39) we obtain

$$G(\vec{r}, \vec{r}') = \left(\frac{2\pi}{k(x - x')}\right)^{1/2} \exp\left\{j\frac{\pi}{4} + j\frac{k}{2}\frac{(y-y')^2}{x-x'}\right\} \int_{-\infty}^{\infty} dE\Psi_E(z)\Psi^*_E(z')e^{j\frac{E}{2k}(x-x')}. \tag{2.43}$$

In all practical problems related to radio wave propagation and scattering in the troposphere, the non-uniformities of the refractive index are localised in the boundary layer of the atmosphere and $\varepsilon_m(z) - 1$ can be approximated by a linear function $2z/a$ with $z \to \infty$. Thus Eq. (2.36) has two solutions: $\chi_E^+(z)$ and $\chi_E^-(z)$ behaving asymptotically as Airy functions $w_2\left(\frac{E}{\mu^2} - \mu z\right)$ and $w_2(E/\mu^2 - \mu z)$, where $\mu = k/m$.

Following Ref. [1], we define the function $w_1(t)$ via a contour integral

$$w_1(t) = \frac{1}{\sqrt{\pi}} \int_\Gamma d\xi \exp\left(-\frac{\xi^3}{3} + t\xi\right) \tag{2.44}$$

where contour Γ comes from infinity along the ray $\arg \xi = -2\pi/3$ to zero and then from zero to infinity $\xi \to +\infty$ along the real axis ξ. The function $w_2(t)$ is defined by a complex conjugate of Eq. (2.44).

The functions $\chi_E^+(z)$ and $\chi_E^-(z)$ describe eigen waves coming to and from infinity. The regular (with $z = 0$) solution for $\Psi_E(z)$is given by the composition

$$\Psi_E(z) = A(E)\left[\chi_E^-(z) - S(E)\,\chi_E^+(z)\right] \tag{2.45}$$

where $A(E) = 1/2\sqrt{\pi\mu}$ is a coefficient defined from normalisation to the delta-function Eqs. (2.40) and (2.41). The term $S(E)$ (normally called the S-matrix in quantum mechanics) is determined by a kind of non-uniformity $\varepsilon_m(z) - 1$ and the boundary condition at $z = 0$. In the case of an ideal boundary condition at $z = 0$, Eq. (2.38), the law of conservation for a number of particles holds true and $|S(E)| = 1$. Therefore, for real E, $S(E) = \exp(-2j\delta(E))$, where $\delta(E)$ is the real phase of scattering.

The S-matrix $S(E)$ has poles E_n in the upper half-space of E and residues in E_n completely define the field in the shadow region. The value of $\Psi_E(z)$ in the pole $E = E_n$ provides a normalised height function of normal wave with number n:

$$\Psi_{E_n}(z) = \chi_{E_n}(z) \equiv \chi_n(z). \tag{2.46}$$

The normal wave $\chi_n(z)$ with complex value of the propagation constant E_n(Im $E_n > 0$) is unlimited at infinite z. This unlimited growth is a consequence of the exponential decay of the field over the x-coordinate, since the field at $z \to \infty$ is created by radiation propagating from points distant to $-\infty$ over the x-coordinate.

Let us calculate integral (2.43) for normal refraction when $\varepsilon_m(z) = 1 + 2z/a$ at $z > 0$ and the eigenfunctions $\Psi_E(z)$of the continuous spectrum (2.45) are defined by

$$\Psi_E(z) = \frac{1}{2\sqrt{\pi\mu}}\left[w_2\left(\frac{E}{\mu^2} - \mu z\right) - \frac{w_2\left(\frac{E}{\mu^2}\right)}{w_1\left(\frac{E}{\mu^2}\right)} w_1\left(\frac{E}{\mu^2} - \mu z\right)\right]. \tag{2.47}$$

One can expect that, in a line-of-sight region, expression (2.43) shall represent a “reflection formula” [1] in accordance with geometric optics, i.e. represent the field as a sum of the direct wave and the wave reflected from the earth’s surface.

Consider ray equations corresponding to the waves radiated either in an upward direction from the source ($z > z_0$) or downward from the source ($z > z_0$), and define the relation between the grazing angle of the ray and stationary value of E. We define the hit-angle ϑ to be a sliding angle of the ray relative to the surface $z > z_0$ at the point of the source location $\vec{r}_0 = \{0, y_0, z_0\}$.

The wave propagating in the direction up from the source can be expressed as follows

$$V^{+} \sim e^{jxE/2k} w_1\left(\frac{E}{\mu^2} - \mu z\right) w_2\left(\frac{E}{\mu^2} - \mu z_0\right). \tag{2.48}$$

Using an asymptotic expression for the Airy function we obtain the trajectory equation by defining the extremum of the phase in Eq. (2.48)

$$z^{+}(x) = \frac{x^2 \mu^3}{4k^2} + \frac{x}{k}\sqrt{\mu^3 z_0 - E} + z_0. \tag{2.49}$$

A similar equation can be obtained for the ray in the direction down from the source

$$z^{-}(x) = \frac{x^2 \mu^3}{4k^2} - \frac{x}{k}\sqrt{\mu^3 z_0 - E} + z_0. \tag{2.50}$$

Differentiating Eqs. (2.49) and (2.50) we can find the equation for angle ϑ and its relation to the stationary value of E:

$$\vartheta = \pm\, a \tan\left(\frac{1}{k}\sqrt{\mu^3 z_0 - E}\right), \tag{2.51}$$

$$E = \mu^3 z_0 - k^2 \tan^2 \vartheta. \tag{2.52}$$

As observed from Eq. (2.51) the values of E satisfying the inequality $E \le \mu^3 z_0$ determine the real hit-angles ϑ, those, in turn, represent uniform plane waves with angle ϑ/m between the normal to their wavefront and tangential to the radius $r = a + z_0$ at the point of source location. Trajectory (2.50) has a minimum in $x_{\min} = \frac{2k}{\mu^3}\sqrt{\mu^3 z_0 - E}$, and $z_{\min}(x_{\min}) = \frac{E}{\mu^3}$. From that it follows that the values $E < 0$ determine the waves reflected from the earth's surface. The sector $-a \tan\left(\sqrt{kz_0/m^3}\right) < \vartheta < a \tan\left(\sqrt{kz_0/m^3}\right)$ determines the "space" rays (direct wave) related to a stationary value of E from the interval

$$0 < E < \mu^3 z_0.$$

The ray trajectories described by Eqs. (2.49) and (2.50) are shown in Figure 2.1.

Therefore, in the line-of-sight region, i.e. $x < \sqrt{2a}\left(\sqrt{z} + \sqrt{z_0}\right)$ we can select two areas of integration in Eq. (2.43):

$$I_1 = \int_{-\infty}^{0} dE \Psi_E(z) \Psi_E^{*}(z_0) e^{j\frac{E}{2k}x}, \tag{2.53}$$

$$I_2 = \int_{0}^{\infty} dE \Psi_E(z) \Psi_E^{*}(z_0) e^{j\frac{E}{2k}x}. \tag{2.54}$$

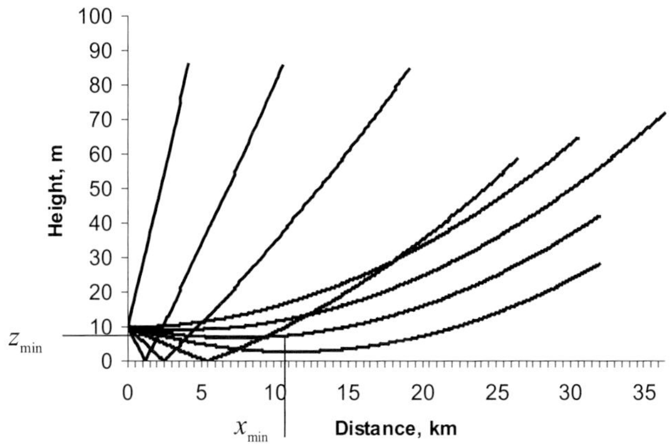

Figure 2.1 Ray trajectories in the case of normal refraction.

Let us substitute Eqs. (2.47)–(2.49) into Eq. (2.53) and assume $z > z_0$ and

$$kz_0/\mathrm{m} >> 1. \tag{2.55}$$

The integration in Eq. (2.53) can be performed using the method of stationary phase similar to the approach in Ref. [1]. The result is given by

$$I_1 = \theta(\sqrt{2a}(\sqrt{z} - \sqrt{z_0}) - x)I_{11} + I_{12}, \tag{2.56}$$

$$I_{11} = 2\mu A^2\sqrt{\frac{2k\pi}{x}}\exp\left\{-j\frac{\pi}{4} + j\frac{k}{2x}(z - z_0)^2 + j\frac{kx}{2a}(z + z_0) - j\frac{1}{12}\left(\frac{mx}{a}\right)^3\right\},$$

$$I_{12} = \sqrt{\frac{k}{a}}\frac{\sqrt{-\xi}}{4\sqrt{\pi}}\frac{1}{\sqrt[4]{zz_0}} \times$$

$$\exp\left\{j\left(\frac{k}{m}\right)^{3/2}\left(z^{3/2} + z_0^{3/2}\right) + j\frac{kx}{2a}(z + z_0) - j\frac{1}{12}\left(\frac{m\xi}{a}\right)^3\right\} \tag{2.57}$$

where $\xi = x - \sqrt{2a}(\sqrt{z} + \sqrt{z_0})$.

Consider integral (2.54) and make use of the asymptotic expression for Airy functions $w_{1,2}\,(E/\mu^2)$ for large positive arguments:

$$w_1\left(\frac{E}{\mu^2}\right) \approx w_2\left(\frac{E}{\mu^2}\right) \approx \left(\frac{E}{\mu^2}\right)^{1/2}\exp\left(\frac{2}{3\mu^3}E^{3/2}\right). \tag{2.58}$$

Substituting Eq. (2.57) into Eq. (2.47) we obtain $S(E) = 1$. Let us introduce the Airy function $v(t)$ by the relationship

$$2jv(t) = w_1(t) - w_2(t) \tag{2.59}$$

and the variable $\tau = E/\mu^2 - \mu z_0$. The integral (2.54) can then be transformed into

$$I_2 \approx \frac{1}{4\pi} e^{j\frac{x}{2k}\mu^3 z_0} \int_{-\mu z_0}^{\infty} e^{j\frac{x\tau}{2k}\mu^2} v(\tau) v(\tau - \mu(z - z_0)) d\tau. \tag{2.60}$$

Without significant error the low limit in the integral (2.60) can be extended to $-\infty$. Substituting into Eq. (2.60) the integral representation for $v(t)$

$$v(t) = \frac{1}{2\sqrt{\pi}} \int_{-\infty}^{\infty} \exp\left(j\frac{\xi^3}{3} + j\xi t\right) d\xi \tag{2.61}$$

and performing the integration, we obtain

$$I_2 \approx \sqrt{\frac{k}{2\pi x}} \exp\left\{-j\frac{\pi}{4} + j\frac{k}{2x}(z - z_0)^2 - j\frac{kx}{2a}(z + z_0) - j\left(\frac{mx}{a}\right)^3\right\}. \tag{2.62}$$

Equation (2.62) describes the direct wave in the range of distances $\sqrt{2a}(\sqrt{z} - \sqrt{z_0}) < x < \sqrt{2a}(\sqrt{z} + \sqrt{z_0})$ and transforms into Eq. (2.55) with $x = \sqrt{2a}(\sqrt{z} - \sqrt{z_0})$. Therefore in a whole line-of-sight region the Green function (2.43) can be represented by superposition of the direct wave and the wave reflected from the earth's surface.

Let us consider the Green function's representation (2.43) in a shadow region, i.e. $x > \sqrt{2a}(\sqrt{z} + \sqrt{z_0})$. In this case the contour of integration over dE can be closed in the upper half-plane and the integral is represented by a residue sum over the poles of $S(E)$. In the particular case of normal refraction and ideal boundary conditions, the poles are given by

$$w_1\left(\frac{E}{\mu^2}\right) = 0 \tag{2.63}$$

and $E_n = \frac{k^2}{m^2}\tau_n e^{j\pi/3}$, $\tau_1 = 2.338, \quad \tau_2 = 4.02, \ \ldots$

As a result, we obtain the Green function representation in terms of the normal wave series

$$G(\vec{r}, \vec{r}') = \sqrt{\frac{2\pi}{k(x - x')}} \exp\left\{j\frac{\pi}{4} + j\frac{k}{2}\frac{(y - y')^2}{x - x'}\right\} \times \sum_{n=1}^{\infty} \exp\left(j\frac{E_n}{2k}(x - x')\right) \frac{w_1\left(\frac{E}{\mu^2} - \mu z\right)}{w_1'\left(\frac{E}{\mu^2}\right)} \frac{w_1\left(\frac{E}{\mu^2} - \mu z'\right)}{w_1'\left(\frac{E}{\mu^2}\right)}. \tag{2.64}$$

In conclusion, the Green function has been built as an expansion over the set of complete and orthogonal eigenfunctions of the continuous spectrum. In the case of normal refraction, such a representation produces final formulas for the field of the point source that are similar to those introduced by Fock [1] on the basis of the classical approach of the contour integral.

It is worthwhile to emphasise the purpose of the above presentation for a Green function. In a problem of multiple scattering the whole path is involved in the scattering process and one needs to have a representation for the field (or Green function) which is equally applicable in the line-of-sight and shadow regions. Also, in the process of obtaining the closed set of equations for statistical moments of the field, the Green function has to possess some properties of orthogonality and completeness which can be achieved by using the above expansion over the set of eigenfunctions of the continuous spectrum.

2.4
Feynman Path Integrals in the Problems of Wave Propagation in Random Media

In this section we introduce an alternative representation for a Green function based on Feynman path integrals. This approach is most suitable for the analysis of the wave propagation and scattering in a line-of-sight region. This section provides some basics for study in an unbounded medium, while later in Section 3 we introduce a path integral representation in the presence of a boundary.

The basic Feynman postulate [45] states that the probability amplitude for the transition of a particle-wave from point $\vec{r}_0$ to point $\vec{r}$ can be represented by contributions made by individual trajectories over which the particle-wave can propagate between points $\vec{r}_0$ and $\vec{r}$. The contribution of each trajectory is proportional to $\exp\{jS/\hbar\}$, where S is a classical action and $\hbar$ is Planck's constant. As is well known, the quantum-mechanical probability amplitude of transition is equivalent to the Green function $G(\vec{r}, \vec{r}_0)$ of the Helmholtz equation written in a parabolic approximation, after the time has been replaced by the x-coordinate along the selected direction of propagation and $m/\hbar$ has been replaced by the wavenumber $k = 2\pi/\lambda$, where m is a particle's mass, k is the wave number of the admitted wave, and λ is the wavelength. Turning to the trajectory continuum, we can write

$$G(\vec{r}, \vec{r}_0) = \int D\vec{\rho}(x) \exp\{jkS[\vec{\rho}(x)]\}, \tag{2.65}$$

where $\vec{r} = \{x, \vec{\rho}\}$, $\vec{r}_0 = \{0, \vec{\rho}_0\}$ are the coordinates of the observation point and source, respectively, $\vec{\rho} = \{y, z\}$, $\vec{\rho} = \{y_0, z_0\}$ are the vectors in the plane orthogonal to the direction of propagation x, and $D\vec{\rho}(x)$ is the differential in the space of continuous trajectories. The action S is given by

$$S[\vec{\rho}(x)] = \int_0^x dx' \left[L_0\left(x', \vec{\rho}\left(x'\right)\right) + \frac{1}{2}\delta\varepsilon\left(x', \vec{\rho}\left(x'\right)\right)\right], \tag{2.66}$$

where $L_0(x, \vec{\rho}(x)) = \frac{1}{2}\left(\frac{d\rho}{dx}\right)^2 + U_0(x, \vec{\rho}(x))$ is an unperturbed (when $\delta\varepsilon = 0$) Lagrangian in the small-angle approximation, $U_0(x, \vec{\rho}(x))$ is an unperturbed potential, $U_0(x, \vec{\rho}(x)) = 1/2\langle\varepsilon(x, \vec{\rho}(x)) - 1\rangle$, the angle brackets $\langle ... \rangle$ here and later denote averaging over the ensemble of dielectric permittivity, $\varepsilon(x, \vec{\rho})$, $\delta\varepsilon(x, \vec{\rho})$ is a random component of $\varepsilon(x, \vec{\rho})$, $\langle\delta\varepsilon\rangle = 0$.

Consider calculation of the first two moments of the Green function (2.65) in the case of an unbounded stationary medium filled with random inhomogeneities of dielectric permeability $\varepsilon(x, \vec{\rho})$. In continuous notation the average Green function

$\langle G(\vec{r},\vec{r}_0)\rangle$ and coherence function $\Gamma\left(x,\vec{\rho}_1,\vec{\rho}_2,\vec{\rho}_0',\vec{\rho}_0''\right) = \left\langle G\left(\vec{r}_1,\vec{r}_0'\right)G^*\left(\vec{r}_2,\vec{r}_0''\right)\right\rangle$ we obtain:

$$\langle G(\vec{r},\vec{r}_0)\rangle = \int D\vec{\rho}(x)\exp\{jkS_0[\vec{\rho}(x)]\}\left\langle \exp\left(j\frac{k}{2}\int_0^x dx'\delta\varepsilon\left(x',\vec{\rho}\left(x'\right)\right)\right)\right\rangle, \tag{2.67}$$

$$\begin{aligned}
&\Gamma\left(x,\vec{\rho}_1,\vec{\rho}_2,\vec{\rho}_0',\vec{\rho}_0''\right) = \left\langle G\left(\vec{r}_1,\vec{r}_0'\right)G^*\left(\vec{r}_2,\vec{r}_0''\right)\right\rangle = \\
&\int D\vec{\rho}(x)\int D\vec{R}(x)\exp\left\{jk\int_0^x dx'\frac{d\vec{R}}{dx'}\frac{d\vec{\rho}}{dx'}\right\}\times \\
&\left\langle \exp\left\{j\frac{k}{2}\int_0^x dx'\left[\delta\varepsilon\left(x',\vec{R}\left(x'\right)+\frac{\vec{\rho}\left(x'\right)}{2}\right)-\delta\varepsilon\left(x',\vec{R}\left(x'\right)-\frac{\vec{\rho}\left(x'\right)}{2}\right)\right]\right\}\right\rangle.
\end{aligned} \tag{2.68}$$

In Eq. (2.68) we have introduced the sum $\vec{R}(x) = 1/2(\vec{\rho}_1(x)+\vec{\rho}_2(x))$ and the difference $\vec{\rho}(x) = \vec{\rho}_1(x)-\vec{\rho}_2(x)$ of the trajectories with the following boundary conditions

$$\vec{R}(x'=0) = \frac{1}{2}\left(\vec{\rho}_0'+\vec{\rho}_0''\right)\ \vec{\rho}(x'=0) = \vec{\rho}_0'-\vec{\rho}_0'',$$

$$\vec{R}(x'=x) = \frac{1}{2}(\vec{\rho}_1+\vec{\rho}_2)\ \vec{\rho}(x'=x) = \vec{\rho}_1-\vec{\rho}_2.$$

We can assume that the fluctuations in $\delta\varepsilon$ are statistically uniform $\langle\delta\varepsilon(x_1,\vec{\rho}_1)\delta\varepsilon(x_2,\vec{\rho}_2)\rangle = B_\varepsilon(x_1-x_2,\vec{\rho}_1-\vec{\rho}_2)$ and have different correlation scales: L_x in direction x and $L_\perp$ in the plane (y,z). If $x >> L$ the integral $\int_0^x dx'\delta\varepsilon\left(x',\vec{\rho}\left(x'\right)\right)$ represents a Gaussian random value in accordance with the central-limit theorem. Hence

$$\left\langle \exp\left(j\frac{k}{2}\int_0^x dx'\delta\varepsilon\left(x',\vec{\rho}\left(x'\right)\right)\right)\right\rangle = \exp\left(-\gamma_s\right), \tag{2.69}$$

$$\begin{aligned}
&\left\langle \exp\left\{j\frac{k}{2}\int_0^x dx'\left[\delta\varepsilon\left(x',\vec{R}\left(x'\right)+\frac{\vec{\rho}\left(x'\right)}{2}\right)-\delta\varepsilon\left(x',\vec{R}\left(x'\right)-\frac{\vec{\rho}\left(x'\right)}{2}\right)\right]\right\}\right\rangle = \\
&\exp\left(-D_s\right),
\end{aligned} \tag{2.70}$$

where

$$\gamma_S = \frac{k^2}{8}\int_0^x dx'\int_0^x dx''\left\langle\delta\varepsilon\left(x',\vec{\rho}\left(x'\right)\right)\delta\varepsilon\left(x'',\vec{\rho}\left(x''\right)\right)\right\rangle \tag{2.71}$$

is the variance of the phase fluctuations along the trajectory $\vec{\rho}(x)$ and

$$D_S = \frac{k^2}{8}\int_0^x dx' \int_0^x dx'' \left\langle \begin{bmatrix} \delta\varepsilon\left(x', \vec{R}(x') + \frac{\vec{\rho}(x')}{2}\right) - \delta\varepsilon\left(x', \vec{R}(x') - \frac{\vec{\rho}(x')}{2}\right) \end{bmatrix} \times \begin{bmatrix} \delta\varepsilon\left(x'', \vec{R}(x'') + \frac{\vec{\rho}(x'')}{2}\right) - \delta\varepsilon\left(x'', \vec{R}(x'') - \frac{\vec{\rho}(x'')}{2}\right) \end{bmatrix} \right\rangle \quad (2.72)$$

is a structure function of the phase difference along the trajectories $\vec{R}(x) + \dfrac{\vec{\rho}(x)}{2}$ and $\vec{R}(x) - \dfrac{\vec{\rho}(x)}{2}$.

Introduce a two-dimensional spatial spectrum of the fluctuations $\delta\varepsilon$ in a plane (y, z):

$$F_\varepsilon\left(x' - x'', \vec{\kappa}_\perp\right) = \frac{1}{4\pi^2}\int d^2\vec{\rho} B_\varepsilon\left(x' - x'', \vec{\rho}\right) e^{-j\vec{\kappa}\vec{\rho}}. \quad (2.73)$$

Then for γ_s and D_s we obtain

$$\gamma_S = \frac{k^2}{4}\int_0^{x/2} d\eta \int_{-2\eta}^{2\eta} d\xi \int d^2\vec{\kappa}_\perp F_\varepsilon(\xi, \vec{\kappa}_\perp) \exp\left(j\vec{\kappa}\xi\frac{d\vec{\rho}}{d\eta}\right), \quad (2.74)$$

$$D_S\left[\vec{R}, \vec{\rho}\right] = \frac{k^2}{4}\int_0^{x/2} d\eta \int_{-2\eta}^{2\eta} d\xi \int d^2\vec{\kappa}_\perp F_\varepsilon(\xi, \vec{\kappa}_\perp) \exp\left(j\vec{\kappa}\xi\frac{d\vec{R}}{d\eta}\right) \times$$

$$\left[\cos\left(\vec{\kappa}\xi\frac{d\vec{\rho}}{d\eta}\right) - \cos\left(\vec{\kappa}\vec{\rho}\right)\right]. \quad (2.75)$$

To analyse Eqs. (2.74) and (2.75) let us introduce the parameter $\theta_R = \max\left[\dfrac{d\vec{R}}{dx}\right]$, the characteristic angle of the ray's trajectory along the *x*-axis. The value of θ_R is contributed to by the three characteristic parameters: θ_d, the gradient of the ray trajectory due to either the position of the correspondent or regular refraction (if any), in case of normal refraction θ_d is given by Eq. (2.51); $\theta_F = 1/\sqrt{kx}$, the angular size of the Fresnel zone; and θ_s, the characteristic angle of scattering on the fluctuations $\delta\varepsilon$, $\theta_s \sim 1/kL_\perp$. Therefore $\theta_R = \max\{\theta_d, \theta_F, \theta_s\}$.

Let us also introduce a parameter of anisotropy α in the fluctuations of $\delta\varepsilon$: $\alpha = L_\perp/L_x$ The effective width of function $F_\varepsilon(\xi, \vec{\kappa}_\perp)$ over ξ does not exceed a correlation scale L_x. When inequality

$$\theta_R << \alpha \quad (2.76)$$

holds, we can assume that $\vec{\kappa}_\perp \xi d\vec{\rho}/d\eta, \vec{\kappa}_\perp \xi d\vec{R}/d\eta << 1$ in a significant region over ξ in Eqs. (2.74) and (2.75). As a result, we obtain the following approximations:

$$\gamma_S(x) = \frac{\pi k^2 x}{4}\int d^2\kappa_\perp \Phi_\varepsilon(0, \kappa_\perp); \quad (2.77)$$

and

$$D_S\left[\vec{R},\vec{\rho}\right] \equiv D_S[\vec{\rho}] = \frac{\pi k^2}{2}\int_0^x dx' \int d^2\kappa_\perp \Phi_\varepsilon(0,\kappa_\perp)\left[1-\cos\left(\kappa_\perp \vec{\rho}(x')\right)\right] \tag{2.78}$$

where

$$2\pi\Phi_\varepsilon(0,\vec{\kappa}_\perp) = \int_{-\infty}^{\infty} d\xi F_\varepsilon(\xi,\vec{\kappa}_\perp) \tag{2.79}$$

and $\Phi_\varepsilon(\vec{\kappa})$ is a three-dimensional spatial spectrum of fluctuations in $\delta\varepsilon$. We can reasonably assume that $\theta_R \sim \theta_s$, and the inequality (2.76) takes the form

$$\frac{kL_\perp^2}{L_x} >> 1 \tag{2.80}$$

which means that the longitudinal correlation scale has to be small compared with the distance $kL_\perp^2$ where the diffraction on the irregularities of the scale $L_\perp$ become significant. The expressions (2.69) and (2.78) provide the fundamental solution to the equation of the coherence function Γ in the Markov approximation:

$$\frac{\partial\Gamma}{\partial x} - \frac{j}{k}\frac{\partial^2\Gamma}{\partial\vec{R}\partial\vec{\rho}} - \frac{\pi k^2}{4}H(\vec{\rho})\Gamma = 0, \tag{2.81}$$

$$H(\vec{\rho}) = 2\int d^2\kappa_\perp \Phi_\varepsilon(0,\kappa_\perp)\left[1-\cos\left(\kappa_\perp\vec{\rho}(x')\right)\right]$$

and $D_S[\vec{\rho}] = \pi k^2/4\int_0^x dx' H\left(\vec{\rho}\left(x'\right)\right)$. "Local" conditions of the Markov approximation are bounded by inequalities (2.67) and (2.77), for an average field we also need smallness of the attenuation of the average field over the distance of L_x: $\gamma_S(L_x) << 1$. "Non-local" conditions were obtained in Ref. [21] and lead to the inequality

$$R_c >> \lambda, \tag{2.82}$$

where R_c is a coherence radius of the scattered field in a plane (y, z), which depends on the distance x, $R_c \equiv R_c(x)$. The parameter R_c can be found from

$$D_s(R_c) = 1. \tag{2.83}$$

To illustrate the path integral approach, we obtain a known solution to Eq. (2.81) for the unbounded medium filled with statistically uniform fluctuations of $\delta\varepsilon(x,\vec{\rho})$. Using Eqs. (2.68), (2.80) and (2.78) we obtain

$$\Gamma\left(x,\vec{\rho}_1,\vec{\rho}_2,\vec{\rho}\,'_0,\vec{\rho}\,''_0\right) = \int D\vec{\rho}(x)\int D\vec{R}(x)\times\exp\left\{jk\int_0^x dx'\frac{d\vec{R}}{dx'}\frac{d\vec{\rho}}{dx'} - D_S\left[\vec{\rho}\left(x'\right)\right]\right\}. \tag{2.84}$$

The integral over trajectories R(x′) represents a continuous Fourier transform of the delta-function $\delta(d^2\vec{\rho}/dx^2)$ [45]. A Lagrangian in the exponent can be expanded into a functional series up to the second order terms in the vicinity of the trajectory

$\vec{\rho}_0\left(x'\right) = \vec{\rho}x'/x + \vec{\rho}_0\left(1 - x'/x\right)$. Integrating then over $\vec{\rho}(x)$, we obtain the well-known solution for an unbounded medium:

$$\Gamma\left(x, \vec{\rho}_1, \vec{\rho}_2, \vec{\rho}'_0, \vec{\rho}''_0\right) = \frac{k^2}{4\pi^2 x^2} \exp\left\{ \begin{array}{l} jk\frac{\left(\vec{R}-\vec{R}_0\right)\left(\vec{\rho}-\vec{\rho}_0\right)}{x} - \\ \frac{\pi k^2}{4}\int_0^x dx' H\left(\vec{\rho}\frac{x'}{x} + \vec{\rho}_0\left(1-\frac{x'}{x}\right)\right) \end{array} \right\}. \tag{2.85}$$

The relationship between the average of the path integrals (2.67), (2.68) and the average over the Fermat paths was analysed in Refs. [34,36]. The asymptotic evaluation of the path integral representation for the Green function (2.67) may be performed using the method of stationary phase. The asymptotic method consists of finding the extreme path $\vec{\rho}^*(x)$ which renders the minimum value of the phase $S[\vec{\rho}(x)]$ given by Eq. (2.66). It can be noted that in obtaining the solution for the coherence function in an unbounded medium (2.85) we actually did not face any difficulties with integration over $R(x)$ due to an isotropy of the medium, $\langle\varepsilon(\vec{r})\rangle = const$. However, this is not the case for more complex media. The variational problem for determining $\vec{\rho}^*(x)$ gives rise to the Euler equation for the Fermat paths

$$\frac{d^2\vec{\rho}}{dx^2} + \nabla\delta\varepsilon(x, \vec{\rho}(x)) = 0, \tag{2.86}$$

where $\nabla = \left\{\frac{\partial}{\partial y}, \frac{\partial}{\partial z}\right\}$.

The rigorous mathematical solution to the problem encountered difficulties in a general case of the multi-scale inhomogeneities of $\delta\varepsilon$ [34] and, therefore, the basic physics has to be involved for a qualitative analysis.

As mentioned above, there are two characteristic transverse scales of the phase fluctuations in $S[\vec{\rho}]$. The first is the Fresnel zone size resulting from the first term in Eq. (2.67). The second term is due to a random component of the dielectric permittivity $\delta\varepsilon$ and is of the order of the coherence scale R_c introduced by Eq. (2.83).

We assume a small value of the mean-square variation of the fluctuations in a phase difference of the fields approaching any points separated in space by a distance of the order of the Fresnel zone size. Hence, this assumption is rendered by the inequality

$$\delta S_1 = \frac{k^2\pi}{4}\int_0^x dx' H\left(\sqrt{\frac{x'}{\kappa}}\right) << 1. \tag{2.87}$$

As known [21], the meaning of this inequality is that the Fresnel zone size is the characteristic region in the plane $x' = $ constant, from which the rays arrive at the receiving point in phase, even in the presence of the $\delta\varepsilon$ fluctuations. It also means that the integration in Eq. (2.67) can be fulfilled along a single canal-ray tube bounded in lateral cross-section by the Fresnel volume. We assume the extreme trajectory $\vec{\rho}^*(x)$ to deviate slightly from the unperturbed path $\vec{\rho}_0(x)$, defined by the unperturbed Lagrangian $L_0[\vec{\rho}(x)]$, therefore, we should require the fluctuations in the arrival angle of the wave defined by $\vec{\rho}^*(x)$ to be small compared with angular size of the Fresnel zone $\sim(kx)^{-1/2}$. In a functional representation this requirement is equivalent to

$$\delta S_2 = \frac{k^2\pi}{8}\int_0^x dx' \delta\vec{\rho}^{\,2}\left(x'\right)\nabla^2 H\left(\vec{\rho}\left(x'\right)\right) \le \frac{k^2\pi}{8} x^2 \int d^2\vec{\kappa}_\perp \Phi_\varepsilon(0, \vec{\kappa}_\perp) << 1. \tag{2.88}$$

In Eq. (2.71) we evaluated the deviation $\delta\vec{\rho} = \vec{\rho}^{\,*} - \vec{\rho}_0$ to be of the order of the Fresnel zone size.

Having satisfied inequalities (2.87) and (2.88), the integration in Eq. (2.67) can be performed along the unperturbed ray trajectories $\vec{\rho}_0(x)$ which are the Fermat paths for a free wave particle propagation (in the particular case of an unbounded medium):

$$\vec{\rho}_0(x') = (\vec{\rho} - \vec{\rho}_0)\frac{x'}{x}. \tag{2.89}$$

Such a straight-line approximation of trajectories in a path integral is similar to the extended Huygens–Fresnel principle introduced in Ref. [46], and used in Ref. [47]. The applicability of extended Huygens–Fresnel principle was analysed in [35], where shown that this approximation besides yielding the exact solution for the first two moments, provides a qualitatively correct solution for the high-order moments of the wave field.

2.5 Numerical Methods of Parabolic Equations

Among many numerical methods used in the problems of applied electromagnetics and wave propagation one may distinguish two methods most widely used in a VHF/UHF propagation in the atmospheric boundary layer, namely: split-step-Fourier and split-step Padé methods. Both methods are based on the parabolic approximation to a wave equation and differ in the method of obtaining the approximation of the exponential operator. Initially, the split-step parabolic approximation was introduced in a problem of applied acoustics [48] where the wave propagation can be described by a scalar wave equation of elliptical type

$$\frac{\partial^2 u}{\partial x^2} + \frac{\partial^2 u}{\partial z^2} + k^2 n^2 u = 0 \tag{2.90}$$

where u is the spectral amplitude of the wave component with frequency ω, $k = \omega/c$ is the wavenumber, c is the group velocity of the propagation and n is the refractive index of the medium and may vary with the coordinates, we assume, for now, a constant value of n, $n \equiv 1$. Equation (2.90) is written for a two-dimensional case of the propagation since this case is the one most widely considered in applications of the split-step approximation.

Equation (2.90) can be factorised as follows

$$\left[\frac{\partial}{\partial x}+j\sqrt{k^2+\frac{\partial^2}{\partial z^2}}\right]\cdot\left[\frac{\partial}{\partial x}-j\sqrt{k^2+\frac{\partial^2}{\partial z^2}}\right]u=0, \tag{2.91}$$

representing the forward and backward propagating waves respectively. We may also remove fast phase variations with distance by introducing the slow-varying amplitude $\tilde{u}$, $u=\tilde{u}\exp(jkx)$. The truncated and factorised equation for amplitude $\tilde{u}$ takes the form

$$\left[\frac{\partial}{\partial x}+jk\left(1+\sqrt{1+\frac{1}{k^2}\frac{\partial^2}{\partial z^2}}\right)\right]\cdot\left[\frac{\partial}{\partial x}+jk\left(1-\sqrt{1+\frac{1}{k^2}\frac{\partial^2}{\partial z^2}}\right)\right]\tilde{u}=0. \tag{2.92}$$

By retaining the only forward propagating wave we obtain the forward parabolic equation

$$\left[\frac{\partial}{\partial x}+jk\left(1-\sqrt{1+\frac{1}{k^2}\frac{\partial^2}{\partial z^2}}\right)\right]\tilde{u}=0 \tag{2.93}$$

which has a formal solution

$$\tilde{u}(x+\Delta x)=\exp\left\{jk\Delta x\left[\sqrt{1+\frac{1}{k^2}\frac{\partial^2}{\partial z^2}}-1\right]\right\}\tilde{u}(x) \tag{2.94}$$

where Δx is a range increment. As apparent from Eq. (2.94) the forward propagating wave at range $x+\Delta x$ can be obtained from values of the wave amplitude at the previous distance x by applying the exponential operator in Eq. (2.94). In order to actually calculate the wave field, the exponential operator in Eq. (2.94) should be approximated in a form suitable for computations. Let us introduce a notation $Z=1/k^2\,\partial^2/\partial z^2$ and, depending on the type of approximation, we obtain three known types of the split-step approximation to parabolic equations:

(a) The standard or narrow angle approximation, obtained by expanding the square root in the exponent into a Taylor series and retaining the linear term:

$$\sqrt{1+Z}=1+\frac{Z}{2}. \tag{2.95}$$

This approximation is used in a split-step Fourier method to be discussed later, and is normally valid for narrow angles of propagation relative to axis x, not exceeding 15°–20°.

(b) The Claerbout approximation in the form

$$\sqrt{1+Z}\approx\frac{1+\frac{3}{4}Z}{1+\frac{Z}{4}} \tag{2.96}$$

which is reported to be valid for wider angles up to 30–40° [49, 50]. And finally:

(c) The split-step Padé approximation, when exponential is expanded into series

$$\exp\left[jk\Delta x\sqrt{1+Z}-1\right] \approx 1 + \sum_{m=1}^{M} \frac{a_m Z}{1+b_m Z} \tag{2.97}$$

where the coefficients a_m, b_m are determined numerically in the complex plane using the approach developed in Ref. [49]. The split-step Padé approximation is reported [50, 51] to be valid for angles of propagation relative to the x-axis of up to 90°.

All the above methods have been realized in a very powerful computational technique well suited for parallel computing and widely used in numerous applications. With regard to the problem of radio wave propagation in the earth's troposphere the most established computational approach is based on the split-step Fourier method equivalent to a standard narrow angle approximation of the square root in the exponential operator. Whilst the wide angle approximations, such as Claerbout and Padé approximations have been used in the problems of scattering on objects submerged in a free-space [50, 52], the author does not know of any systematic derivation of the Padé approximation in the case of radio wave propagation in a stratified troposphere over terrain or the sea boundary surface. We may notice that while the fundamental elliptic equation of type (2.90) can be obtained for potentials (Hertz functions) under free-space propagation conditions, the equations for the Debye's potentials take a different form in the case of a stratified troposphere [1]. In the presence of a random component of the refractive index the main equation for the slow varying envelope is given by Eq. (2.26).

Consider Eq. (2.26) and retain the term $\frac{1}{r^2}\frac{\partial^2 W_1}{\partial \theta^2}$ on the right-hand side of the equation. While this is not very consistent with the arguments used in Section 2.1, we retain this term in order to get in line with the approach of other authors to obtaining the parabolic equation. We may notice that to strictly follow the procedure we have to leave the other terms in the right-hand side of Eq. (2.26) in order to be consistent in the accuracy of the approximation. Introducing the same coordinates $x = a\theta$, $z = r - a$ and $y = \phi \cdot a \sin\vartheta$ we obtain

$$\frac{\partial^2 W_1}{\partial x^2} + j2k\frac{\partial W_1}{\partial x} + \Delta_\perp W_1 + k^2\left[(\varepsilon_m(z)-1) + \delta\varepsilon(\vec{r})\right]W_1 = 0. \tag{2.98}$$

Then we may introduce the operator

$$L = \frac{1}{k^2}\left[\Delta_\perp + k^2\left[(\varepsilon_m(z)-1) + \delta\varepsilon(\vec{r})\right]\right] \tag{2.99}$$

and factorize Eq. (2.98) in a form similar to Eq. (2.90),

$$\left[\frac{\partial}{\partial x} + jk\left(1+\sqrt{1+L}\right)\right]\cdot\left[\frac{\partial}{\partial x} + jk\left(1-\sqrt{1+L}\right)\right]W_1 = 0.$$

Then, following the above described procedure, we may retain the forward propagating wave obeying the truncated equation

$$\frac{\partial W_1}{\partial x} = jk\left(\sqrt{1+L} - 1\right) W_1 \tag{2.100}$$

with formal solution at the marching step in the form

$$W_1(x+\Delta x) = W_1(x)\exp\left[jk\Delta x\left(\sqrt{1+L} - 1\right)\right]. \tag{2.101}$$

The exponential can then be expanded into a series similar to Eq. (2.97) and the next range step solution is given by

$$W_1(x+\Delta x, y, z) = W_1(x) + \sum_{m=1}^{M} \frac{a_m L}{1+b_m L} W_1(x, y, z). \tag{2.102}$$

Following the approach in Ref. [50], we may introduce the auxiliary function

$$f_m(x+\Delta x, y, z) = \frac{a_m L}{1+b_m L} W_1(x, y, z) \tag{2.103}$$

which can be found by multiplying both parts of Eq. (2.103) by $1 + b_m L$ and solving a second-order differential equation for f_m. Substituting f_m into Eq. (2.102) we have

$$W_1(x+\Delta x, y, z) = W_1(x) + \sum_{m=1}^{M} f_m(x+\Delta x, y, z). \tag{2.103}$$

It is important to notice that all the functions f_m can be solved independently and in parallel at each step along the distance x. The marching equation (2.103), as well as the forward propagating wave equation (2.93), shall be appended by both boundary conditions (at $z = 0$ and $|\vec{r}| \to \infty$) and initial conditions at $x = 0$. A comprehensive treatment of the Padé series can be found in Ref. [51].

A somewhat modified approach to the implementation of the Padé method is to substitute a Padé series with a Padé product

$$\sum_{m=1}^{M} \frac{a_m L}{1+b_m L} = \prod_{m=1}^{M} \frac{1+\lambda_m L}{1+\mu_m L}. \tag{2.104}$$

This approach has been developed in Refs. [48, 53] and was shown to be well suited for finite-difference computational schemes.

Now, we concentrate on the split-step Fourier method. Consider Eq. (2.26) and assume the presence of deterministic and stratified refractivity in the medium, the random component $\delta\varepsilon$ is set to 0. Assume first that the propagation takes place in an unbounded medium, i.e., we have only requirements on the appropriate decay of the field at infinity. The next assumption normally employed in the split-step method applied to radio wave propagation in the troposphere is to assume the medium to be stratified over the z-coordinate, i.e. a stratified troposphere, thus removing the dependence on the y-coordinate and truncating Eq. (2.26) to a two-dimensional (x, z) parabolic equation.

Equation (2.26) in this case can be presented in the form

$$\frac{\partial W}{\partial x} = j\left[\frac{k}{2}(\varepsilon_m(z) - 1) + \frac{1}{2k}\frac{\partial^2}{\partial z^2}\right]W. \tag{2.105}$$

In a vertically stratified medium the formal solution to Eq. (2.105) is then given by

$$W(x + \Delta x) = \exp\left\{j\Delta x\left[\frac{k}{2}(\varepsilon_m(z) - 1) + \frac{1}{2k}\frac{\partial^2}{\partial z^2}\right]\right\}W(x) \tag{2.106}$$

where Δx is a range increment. Equation (2.106) forms the basis for a split-step Fourier implementation.

We follow the basics of the split-step Fourier implementation reported in Ref. [54]. Let us define a Fourier transform of the envelope $W(x, z)$ as follows:

$$\tilde{W}(x, p) = \Im[W(x, z)] = \int_{-\infty}^{\infty} dz W(x, z)\exp(-jpz). \tag{2.107}$$

Apparently p is a vertical component of the wave vector of the incident field $W(x, z)$.

The fundamental assumption in a split-step solution to Eq. (2.106) is that a Fourier transform of the envelope $W(x, z)$ is performed while treating the term $\frac{k}{2}(\varepsilon_m(z) - 1)$ as a constant, i.e., no dependence on the z-coordinate. With that assumption, we obtain a marching solution

$$\tilde{W}(x + \Delta x, p) = \tilde{W}(x, p)\exp\left[-\left(k^2(\varepsilon_m(z) - 1) - p^2\right)\frac{\Delta x}{2jk}\right]. \tag{2.108}$$

The inverse Fourier transform is given by

$$W(x + \Delta x, z) = \exp\left[j\frac{k}{2}(\varepsilon_m(z) - 1)\Delta x\right] \cdot \Im^{-1}\left\{\tilde{W}(x, p)\exp\left(-jp^2\frac{\Delta x}{2k}\right)\right\}. \tag{2.109}$$

In the above equation the term $\frac{k}{2}(\varepsilon_m(z) - 1)$ is no longer a constant, its variations actually results in variation of the angular spectrum p of the marching field $W(x + \Delta x, z)$ with the distance.

The boundary condition in the case of ideal reflection (a perfectly conducting boundary surface) can be realized by adding the mirror image of the field $W(x, z)$ in the upper half-space ($z > 0$) into the lower half space ($z < 0$) with an appropriate phase to reproduce the odd or even image of the field W relative to $z = 0$. The odd composition satisfies a condition $W(x, z = 0) = 0$ for a vertically polarized field, while the even composition results in the boundary condition $\partial W/\partial z|_{z=0}$ for horizontal polarization. In these cases of perfectly conducting boundary the Fourier transform reduces to a one-sided sine or cosine transform, respectively

$$\tilde{W}_o(x, p) = -2j\Im[W_o(x, z)] = -2j\int_0^{\infty} dz W_o(x, z)\sin(pz),$$

$$\tilde{W}_e(x,p) = 2\Im[W_e(x,z)] = 2\int_0^\infty dz W_e(x,z)\cos(pz) \tag{2.110}$$

where the subscripts e and o indicate that the field is even or odd, which, in turn, corresponds to horizontal or vertical polarization, respectively.

The impedance boundary condition (2.27) was also treated in Ref. [54] where it was shown that such a boundary condition can be realized using a mixed Fourier transform

$$\tilde{W}_q(x,p) = \int_0^\infty dz W(x,z)[q\sin(pz) - p\cos(pz)]$$

where $q = \frac{jm}{\sqrt{\eta+1}}$ (see Section 2.2). The marching solution for $W(x, z)$ is then given by [54]:

$$W(x+\Delta x, z) = \exp\left[j\frac{k}{2}(\varepsilon_m(z)-1)\Delta x\right]\times$$

$$\left\{\begin{array}{l} \exp\left(jq^2\frac{\Delta x}{2k} - qz\right)K(x) + \frac{2}{\pi}\times \\ \int_0^\infty dp\frac{q\sin(pz)-p\cos(pz)}{q^2+p^2}\exp(-j\frac{p^2}{2k}\Delta x)\times \\ \int_0^\infty dz' W(x,z')\left(q\sin(pz') - p\cos(pz')\right) \end{array}\right\}$$

where the term $K(x)$ is defined as [54]:

$$K(x) = 2q\int_0^\infty dz W(x,z)\exp(-qz); \qquad \mathrm{Re}(q) > 0, \tag{2.112}$$

$$K(x) = 0; \qquad \mathrm{Re}(q) < 0.$$

It is apparent that the implementation of the split-step Fourier algorithm employing the formulas (2.111) and (2.112) will be more complicated than the rather simple implementation for a perfectly conducting interface using formula (2.110). As shown in Ref. [54] and many other publications (we may refer to a practical implementation of the numerical methods in radio coverage prediction systems [39, 55, 56]), the value of the implementation of the impedance boundary conditions becomes pronounced at vertical polarization at smaller distances and at lower VHF frequencies for long-range propagation, where the impact of the surface wave produced by the distributed images of the source provides a substantial contribution to the received field strength.

The important development in an implementation of the split step Fourier transform method is the capability to treat an irregular terrain profile, as reported in numerous publications (see, e.g. Refs. [55, 57–59]. The approach was initially developed in underwater acoustics [60] and then implemented in the studies of radio propagation in the troposphere [55, 57].

Following Ref. [55], let us consider the field of horizontal polarization over an irregular terrain with the terrain profile given by the function of the distance $T(x)$. (We may note that with $|q| >> 1$ the boundary condition is $W(x, z = 0) = 0$ for both polarizations.) The range dependent boundary condition is then given by $W(x, z = T(x)) = 0$. The approach is then to map the irregular terrain profile to a smooth or more flat one with consequent modifications of the wave equation. The mapping is made by introducing a change of variables. Let the new height and range variables be represented by

$$\chi = x, \quad \varsigma = z - T(x). \tag{2.113}$$

In the new coordinate system the slow varying envelope $W(x, z)$ can be written in the form

$$W(x, z) = w(\chi, \varsigma)\exp\left(j\theta(\chi, \varsigma)\right) \tag{2.114}$$

where $\theta(\chi, \varsigma)$ is a phase correction term due to the irregular terrain. Substituting the new variables into Eq. (2.26) and following the procedure described in Ref. [55] we end up with the following equation for a new envelope

$$\frac{\partial^2 w}{\partial \varsigma^2} + j2k\frac{\partial w}{\partial \chi} + k^2\left[(\varepsilon_m(\varsigma + T(\chi)) - 1) + 2\varsigma T''(\chi)\right]w = 0 \tag{2.115}$$

and

$$\theta(\chi, \varsigma) = -k\varsigma T'(\chi) - k^{3/2}\int_0^{\chi}\left[T'(\alpha)\right]^2 d\alpha. \tag{2.116}$$

As observed, the implementation of the split-step Fourier transform can be done in a way similar to the case of a smooth terrain, the difference from Eq. (2.26) is the presence of the second derivative of the boundary interface, $2\varsigma T''(\chi)$, in the exponential operator. Nonetheless, given that the split-step Fourier transform algorithm treats this term as a constant, the addition of the second derivative does not make a difference in the implementation though ensuring the important correction of the wave front due to scattering on the irregular terrain. Specific issues in computer-based implementation of the split-step Fourier algorithm are discussed in Refs. [55, 56].

We may notice apparent similarities between the split-step Fourier method and the discrete implementation of the path integral method [36, 45]. In the case of propagation over the boundary interface, considered in Section 3, implementation of the boundary conditions into the path integral approach will result in the appearance of the mirrored source, similar to the above formulas (2.110) and (2.111). In fact, numerical calculation of the discrete version of the path integral is practically realised by applying the fast-fourier transform algorithm to the exponential propagator at given step along the range x.

Finally, we can state that the above numerical methods constitute a powerful framework for quantitative analysis of radiowave propagation through the tropospheric boundary layer in the very general case of the refractivity condition. In particular, the methods described above can incorporate either non-uniformity of the refractivity profiles in the horizontal plane as well as in an irregular terrain. The limitation of the above methods, as in fact of all numerical methods, is that while they are quite helpful in a quantitative estimation of the propagation phenomena when the physical mechanism is clearly understood they lack the capability to provide a framework for qualitative analysis, especially in the case of combined effects of refraction and scattering on random irregularities of refractive index or a randomly rough sea surface.

2.6
Basics of Fock's Theory

In this section we briefly reproduce the basic results obtained by Fock in Chapter 12 of Ref. [1] for the vertical dipole in the case of normal refraction, in order to have a reference model for the further studies described in this book.

Following Fock [1], we introduce a non-dimensional distance and height ξ, h, respectively:

$$\xi = \frac{mx}{a}, \quad h = \frac{kz}{m} \tag{2.117}$$

where $x = a\theta$ and $z = r - a$, the same physical coordinates as defined in Section 2.2. The relation between the Debye's potential U and the slowly varying attenuation factor (envelope) V is given by

$$U = \frac{\exp(jka\theta)}{\sqrt{a^2\theta\sin\theta}} V(\xi, h_1, h_2, q) \tag{2.118}$$

where h_1 and h_2 are the non-dimensional heights of the receiver and transmitter, respectively, above the earth's surface, ξ is the distance between them along the earth's surface. A normalised impedance q is defined by

$$q = \frac{jm}{\sqrt{\eta+1}}. \tag{2.119}$$

The attenuation factor V is given by a contour integral in a plane of complex parameter t:

$$V(\xi, h_1, h_2, q) = \exp\left(-j\frac{\pi}{4}\right)\sqrt{\frac{\xi}{\pi}} \int_C dt \exp(j\xi t) F(t, h_1, h_2, q). \tag{2.120}$$

The contour C comes from infinity in the second quadrant of the t-plane passing around and below all the poles of the integrand and then goes to infinity in the first quadrant of the t-plane. In the case of $h_2 > h_1$, the function F is given by

$$F(t, h_1, h_2, q) = \frac{j}{2} w_1(t - h_2)\left[w_2(t - h_1) - \frac{w_2'(t) - q w_2(t)}{w_1'(t) - q w_1(t)} w_1(t - h_1)\right]. \tag{2.121}$$

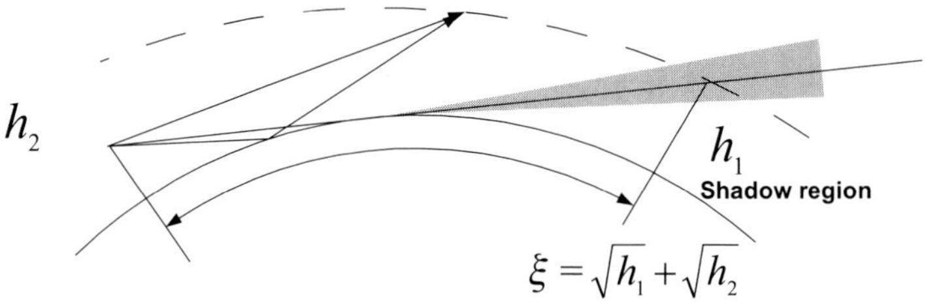

Figure 2.2 Geometry of wave propagation.

In the case $h_1 > h_2$ we need to swap the h_1 and h_2 in the expression for F above. As shown in Ref. [1], three characteristic regions can be selected in the space of ξ, h_1, h_2 or, over the distance ξ, given fixed values of h_1, h_2. The first region is a line-of-sight region with $\xi < \sqrt{h_1} + \sqrt{h_2}$, where the field is composed by superposition of the direct wave and the wave reflected from the earth's surface. The second region is a shadow region at $\xi > \sqrt{h_1} + \sqrt{h_2}$, where the electromagnetic field propagates only due to a diffraction mechanism and, finally, the third region is a transition region with $\xi \approx \sqrt{h_1} + \sqrt{h_2}$, which separates the line-of-sight region from the shadow region. These regions are shown schematically in Figure 2.2. The transition region, also called a "shade cone", is filled with dots in the figure. Fock obtained an asymptotical expansion of the integral (2.120) for all three regions [1]. We reproduce his results for both the line-of-sight and the shadow regions.

In the shadow region the function F can be written in the following form:

$$F(t, h_1, h_2, q) = w_1(t - h_2)\frac{f(h_1, t)}{w_1'(t) - qw_1(t)} \tag{2.122}$$

where

$$f(h_1, t) = \left[w_1'(t) - qw_1(t)\right]v(t - h_1) - \left[v'(t) - qv(t)\right]w_1(t - h_1). \tag{2.123}$$

With $h_1 = 0$ the function f and its derivative have the values

$$f(0, t) = 0,\ df/dh_1 = -q. \tag{2.124}$$

The above equations can be obtained using the representation the Airy function w_1 via the Airy functions u and v, see Appendix 1. From Eq. (2.124) one can find that if $t = t_n$ is a root of the equation

$$w_1'(t) - qw_1(t) = 0,\ t = t_1, t_2, \ldots \tag{2.125}$$

the value of the function f is given by

$$f(h_1, t_s) = f_s(h_1) = \frac{w_1(t_s - h_1)}{w_1(t_s)} \tag{2.126}$$

and can be regarded as a height gain function of the mode with number s. In two limiting cases $q = 0$ and $q = \infty$ the roots t_s can be estimated from asymptotic expressions valid for large s:

$$t_{s,q=0} \cong \left[\frac{3\pi}{2}\left(s-\frac{3}{4}\right)\right]^{2/3} \exp\left(j\frac{\pi}{3}\right),$$
$$t_{s,q=\infty} \cong \left[\frac{3\pi}{2}\left(s-\frac{1}{4}\right)\right]^{2/3} \exp\left(j\frac{\pi}{3}\right), \tag{2.127}$$

and the first roots are as follows: $t_{1,q=0}=1.01879e^{j\pi/3}$; $t_{1,q=\infty}=2.33811e^{j\pi/3}$;

The integral (2.120) can then be calculated as a sum of the residues in the poles of the integrand given by Eq. (2.125):

$$V(\xi,h_1,h_2,q) = \exp\left(j\frac{\pi}{4}\right)2\sqrt{\pi\xi}\sum_{s=1}^{\infty}\frac{e^{j\xi t}}{1-\frac{t_s}{q^2}}\frac{w_1(t_s-h_1)}{w'(t_s)}\frac{w_1(t_s-h_2)}{w'(t_s)}. \tag{2.127}$$

Equation (2.127) represents the series of the normal modes of the diffracted waves converging rapidly in a shadow region.

Consider the field in the line-of-sight interference region, $\xi < \sqrt{h_1} + \sqrt{h_2}$. In that region we have to obtain the reflection formula corresponding to the reflection from the spherical surface. The integral (2.120) can be represented by the sum of two terms

$$V(\xi,h_1,h_2,q) = V_d - V_r,$$

$$V_d = \frac{1}{2}\exp\left(j\frac{\pi}{4}\right)\sqrt{\frac{\xi}{\pi}}\int_C dt \exp(j\xi t) w_1(t-h_2) w_2(t-h_1), \tag{2.128}$$

$$V_r = \frac{1}{2}\exp\left(j\frac{\pi}{4}\right)\sqrt{\frac{\xi}{\pi}}\int_C dt \exp(j\xi t)\frac{w_2'(t)-qw_2(t)}{w_1'(t)-qw_1(t)} w_1(t-h_1) w_1(t-h_2). \tag{2.129}$$

Assuming that the major contribution to integrals (2.128) and (2.129) comes from the interval of large and negative t we may use the asymptotic formulas (A.21), (A.22) for the Airy functions w_1 and w_2. Then for direct wave V_d we obtain

$$V_d = \frac{1}{2}\exp\left(j\frac{\pi}{4}\right)\sqrt{\frac{\xi}{\pi}}\int_C dt \frac{\exp(j\varphi(t))}{\left[(h_1-t)(h_2-t)\right]^{1/4}}. \tag{2.130}$$

The phase of the integrand $\varphi(t)$ is given by

$$\varphi(t) = \xi t + \frac{2}{3}(h_2-t)^{3/2} - \frac{2}{3}(h_1-t)^{3/2}.$$

The stationary value of t_s is determined from $\varphi'(t_s) = 0$. The stationary value of the phase we denote as $\varphi \equiv \varphi(t_s)$ and it is given by

$$\varphi = \frac{(h_1-h_2)^2}{4\xi} + \frac{1}{2}\xi(h_1+h_2) - \frac{\xi^3}{12}. \tag{2.131}$$

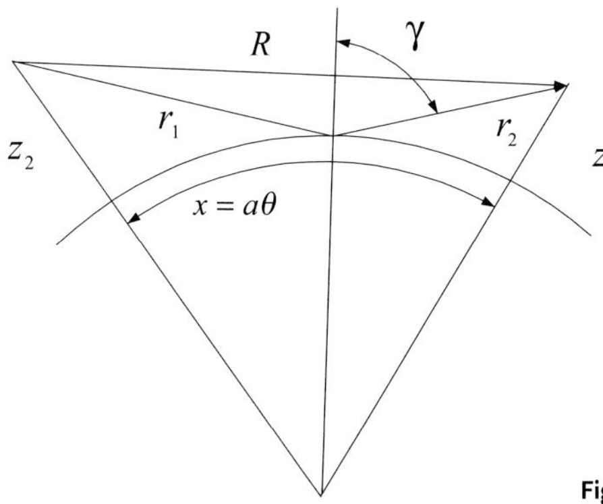

Figure 2.3 Ray traces in the line-of-sight region.

The value of φ has a simple geometric meaning of the difference between the phase of the direct wave along the distance R between the correspondents and the distance $x = a\theta$ along the earth's surface, Figure 2.3:

$$\varphi = k(R - x). \tag{2.132}$$

Application of the stationary phase method to integral (2.130) leads to the following expression for the direct wave

$$V_d = \exp(j\varphi). \tag{2.133}$$

Consider the term V_r. Using a similar asymptotic expression for the Airy function in the integrand we obtain

$$V_d = \frac{1}{2}\exp\left(j\frac{\pi}{4}\right)\sqrt{\frac{\xi}{\pi}}\int_C dt \frac{\exp(j\psi(t))}{[(h_1-t)(h_2-t)]^{1/4}\dfrac{q-j\sqrt{-t}}{q+j\sqrt{-t}}} \tag{2.134}$$

and

$$\psi(t) = \xi t + \frac{2}{3}(h_2 - t)^{3/2} + \frac{2}{3}(h_1 - t)^{3/2} - \frac{4}{3}(-t)^{3/2}. \tag{2.135}$$

The root of the equation $\psi'(t) = 0$ we define as $t = -p^2$, where $p > 0$. The stationary point in terms of p will be the root of the equation

$$\sqrt{h_1 + p^2} + \sqrt{h_2 + p^2} = 2p + x. \tag{2.136}$$

The stationary value of the phase can then be defined in terms of the parameter p:

$$\psi = -3p^2\xi + 2p\left(h_1 + h_2 - \xi^2\right) + \xi(h_1 + h_2) - \frac{\xi^3}{3}. \tag{2.137}$$

The application of the stationary phase method to Eq. (2.134) then results in the following expression for reflected wave:

$$V_r = \frac{q - jp}{q + jp}\sqrt{A}\exp(j\psi) \tag{2.138}$$

with

$$A = \frac{px}{3px + x^2 - h_1 - h_2}. \tag{2.139}$$

The formula (2.138) has the following geometric meaning. The parameter p is $p = m\cos(\gamma)$, where γ is an incidence angle as shown in Figure 2.3. The factor $q - jp/q + jp$ is a Fresnel coefficient of reflection taken with the opposite sign. The parameter $\sqrt{A}$ is a correction term accounting for divergence of the beam after reflection multiplied by a facto R/r_1, where r_1 is the distance along the ray from the source to the reflection point, Figure 2.3. The phase ψ can be written in a form similar to Eq. (2.132):

$$\psi = k(r_1 + r_2 - x). \tag{2.140}$$

Combining both terms in V we obtain the reflection formula

$$V = \exp(j\varphi) - \frac{q - jp}{q + jp}\sqrt{A}\exp(j\psi). \tag{2.141}$$

The actual stationary value of p is given by

$$p = \frac{1}{2x}\left(h_1 + h_2 - \frac{1}{2}\xi^2 + 4\rho^2\sin^2\left(\frac{\alpha}{3}\right)\right) \tag{2.142}$$

where

$$\rho^2 = \frac{1}{3}\left(\xi^2 + 2h_1 + 2h_2\right); \quad \sin(\alpha) = \frac{\xi(h_1 - h_2)}{\rho^2}; \; -\frac{\pi}{2} < \alpha < \frac{\pi}{2}. \tag{2.143}$$

The reflection formula (2.141) is valid for large enough p, practically it gives a good approximation for $p > 2$.

As will be discussed further, experimental results at frequencies above 1 GHz suggest little difference in the propagation of the vertically and horizontally polarised fields. As may be seen from the definition and values of m and η, parameter q is actually large for sea water at frequencies above 1 GHz, which suggests that, at least in the case of long range propagation, i.e. at distances beyond the optical horizon, the impedance boundary conditions can be approximated by conditions for a vertically polarised field with $q \to \infty$, practically suitable for both polarizations.

2.7 Fock's Theory of the Evaporation Duct

In the case of the super-refraction associated with the evaporation duct the M-profile has one minimum at some height Z_s above the sea surface, called the height of the evaporation duct. Considering the case where $q = \infty$, the attenuation factor $V(\xi, h_1, h_2, \infty)$ is given by Eq. (2.120) with the integrand $F(t, h_1, h_2, \infty)$ given by

$$F(t, h_1, h_2) = \frac{j}{2} f_1(t, h_2) \left[f_2(t, h_1) - \frac{f_2(0,t)}{f_1(0,t)} f_1(t, h_1) \right]. \tag{2.144}$$

and the height-gain functions f_1, f_2 are the solutions to the equation

$$\frac{d^2 f}{dh^2} + U(h) f = tf. \tag{2.145}$$

The function $U(h)$ is assumed to be an analytical function of its argument. We concentrate further on h and t where the equation $U(h) - t = 0$ has two roots: $h = b_1$ and $h = b_2$. For real t in the interval $(U(H_s) < t \leq U(0))$, the above roots are also real.

Consider an asymptotical integration of Eq. (2.145) for height-gain functions which is valid for all values h and t, including $t = U(H_s)$, i.e. the point of the minimum of the U-profile. The asymptotical solution in this case should be built upon an "etalon" equation, which behaves similarly to Eq. (2.145), i.e. has a single minimum, and, on the other hand, has a known analytical solution. The "etalon" equation can be expressed in the form of an equation for functions of a parabolic cylinder

$$\frac{d^2 g}{d\varsigma^2} + \left[\frac{1}{4} \varsigma^2 + \nu \right] g = 0. \tag{2.146}$$

The relation between Eqs. (2.146) and (2.145) is established by a transformation of h into ς in a form which ensures that two conditions are met: (a) the parameter $U(h) - t$ in Eq. (2.145) has the same roots as $1/4\varsigma^2 + \nu$, and (b) with large values of the above parameters, either Eq. (2.145) or Eq. (2.146) results in the same asymptotic expression for their respective solutions f and g. Such conditions are satisfied with the following substitution

$$\int_{b_1}^{h} \sqrt{U(h) - t}\, dh = \frac{1}{2} \int_{-2j\sqrt{\nu}}^{\varsigma} \sqrt{\varsigma^2 + 4\nu}\, d\varsigma \tag{2.147}$$

with parameter ν bounded by following equation

$$\int_{b_1}^{b_2} \sqrt{U(h) - t}\, dh = \frac{1}{2} \int_{-2j\sqrt{\nu}}^{2j\sqrt{\nu}} \sqrt{\varsigma^2 + 4\nu}\, d\varsigma. \tag{2.148}$$

The value integral in Eq. (2.146) results in the relationship between roots b_1, b_2, and ν

$$j\pi\nu = \int_{b_1}^{b_2} \sqrt{U(h) - t}\, dh. \tag{2.149}$$

Expanding the right-hand side of Eq. (2.149) into a Taylor series in the vicinity of the minimum of $U(h)$ we obtain ν as a holomorphic function of t

$$\nu = \frac{U(H_s)-t}{\sqrt{2U''(H_s)}} + \ldots \qquad (2.150)$$

Since $U''(H_s) > 0$, we can read formula (2.149) in the following way: $\nu > 0$ when $t < U(H_s)$ and $\nu < 0$ when $t > U(H_s)$; with $U(H_s) < t < U(0)$ we then have

$$\nu = -\frac{1}{\pi}\int_{b_1}^{b_2} \sqrt{t - U(h)}dh. \qquad (2.151)$$

Then, introducing the notations for the phase terms

$$S = \int_0^h \sqrt{U(h) - t}dh, \; S_0 = \frac{1}{2}\int_0^{b_1} \sqrt{U(h) - t}dh + \frac{1}{2}\int_0^{b_2} \sqrt{U(h) - t}dh, \qquad (2.152)$$

the substitution (2.147), which in fact defines the relation between h and ς, can be expressed in the form

$$S(h) - S_0 = \frac{1}{2}\int_{-2j\sqrt{\nu}}^{\varsigma} \sqrt{\varsigma^2 + 4\nu}d\varsigma = \frac{1}{4}\varsigma\sqrt{\varsigma^2 + 4\nu} + \nu \ln\left(\varsigma + \sqrt{\varsigma^2 + 4\nu}\right) - \frac{\nu}{2}\ln(4\nu).$$

(2.153)

The respective solutions of Eqs. (2.145) and (2.146) are related as follows

$$f = \sqrt{\frac{dh}{d\varsigma}}g \qquad (2.154)$$

where function g, the solution to Eq. (2.146), can be expressed via a function of a parabolic cylinder $D_n(z)$, given by the series

$$D_n(z) = \frac{2^{-\frac{n}{2}-1}}{\Gamma(-n)} e^{-\frac{z^2}{4}} \sum_{m=0}^{\infty} \frac{\Gamma\left(\frac{m-n}{2}\right)}{\Gamma(m+1)} 2^{\frac{m}{2}} (-z)^m. \qquad (2.155)$$

Function $D_n(z)$ will obey Eq. (2.146) when $n = j\nu - 1/2$ and $z = \varsigma e^{-j\pi/4}$. The appropriate solutions to Eq. (2.145) are then given by

$$f_1(h,t) = c_1(\nu)\sqrt{2\frac{dh}{d\varsigma}}g_1(\varsigma),$$
$$f_2(h,t) = c_2(\nu)\sqrt{2\frac{dh}{d\varsigma}}g_2(\varsigma) \qquad (2.156)$$

and

$$g_1(\varsigma) = D_{j\nu-1/2}\left(\varsigma e^{-j\pi/4}\right),$$
$$g_2(\varsigma) = D_{-j\nu-1/2}\left(\varsigma e^{j\pi/4}\right), \qquad (2.157)$$

$$c_1(\nu) = \exp\left[-\frac{\pi\nu}{4} + j\frac{\pi}{8} + j\left(\frac{1}{2}\nu - \frac{1}{2}\nu\ln(\nu) - S_0\right)\right],$$
$$c_2(\nu) = c_1(\nu)^* = \exp\left[-\frac{\pi\nu}{4} - j\frac{\pi}{8} - j\left(\frac{1}{2}\nu - \frac{1}{2}\nu\ln(\nu) - S_0\right)\right]. \quad (2.158)$$

The coefficients $c_1(\nu)$ and $c_2(\nu)$ are defined by Eq. (2.158) with the aim being to obtain a correct asymptotic expression for $f_1(h,t)$, $f_2(h,t)$ with large values of the argument h and holomorphic functions (2.156) in the vicinity of $\nu = 0$.

With large positive ς, that corresponds to large heights h, well above H_s, the height gain functions take the following asymptotic forms:

$$f_1(h,t) = \frac{e^{j\pi/4}}{\sqrt[4]{U(h)-t}}\exp(j(S-2S_0)),$$

$$f_2(h,t) = \frac{e^{-j\pi/4}}{\sqrt[4]{U(h)-t}}\exp(-j(S-2S_0)). \quad (2.159)$$

Inside the evaporation duct and, strictly speaking, far below the inversion height H_s when ς is large and negative, the asymptotic representations for the height-gain functions take the form:

$$f_1(h,t) = \chi_1(\nu)\frac{e^{j\pi/4}}{\sqrt[4]{U(h)-t}}\exp(j(S-2S_0)) + e^{-\pi\nu}\frac{e^{-j\pi/4}}{\sqrt[4]{U(h)-t}}\exp(-jS),$$
$$f_2(h,t) = \chi_2(\nu)\frac{e^{-j\pi/4}}{\sqrt[4]{U(h)-t}}\exp(-j(S-2S_0)) + e^{-\pi\nu}\frac{e^{j\pi/4}}{\sqrt[4]{U(h)-t}}\exp(jS) \quad (2.160)$$

where

$$\chi_1(\nu) = \frac{\sqrt{2\pi}}{\Gamma\left(\frac{1}{2} - j\nu\right)}\exp\left[-\frac{\pi\nu}{2} + j(\nu - \nu\ln(\nu))\right],$$
$$\chi_2(\nu) = \frac{\sqrt{2\pi}}{\Gamma\left(\frac{1}{2} + j\nu\right)}\exp\left[-\frac{\pi\nu}{2} - j(\nu - \nu\ln(\nu))\right]. \quad (2.161)$$

The asymptotic representation of the integrand (2.144) is then given by

$$F(t,h_1,h_2) = \frac{j}{2}\frac{1}{\sqrt[4]{U(h_1)-t}\cdot\sqrt[4]{U(h_2)-t}} \times$$
$$\frac{\left[e^{jS(h_1)} - \Lambda e^{-jS(h)}\right]\cdot\left[e^{-jS(h_2)} - e^{jS(h_2)}\right]}{1-\Lambda} \quad (2.162)$$

with $H_s > h_1 > h_2$, where

$$\Lambda = j\frac{e^{-\pi\nu + 2jS_0}}{\chi_1(\nu)}. \quad (2.163)$$

The poles in the integral (2.120) with the integrand in the form (2.162) will be given by the roots of the characteristic equation

$$1 - \Lambda = 0. \tag{2.164}$$

In many practical cases the evaporation duct parameters are such that the values of parameter ς are of the order of 1, or even less, in Eqs. (2.162) and (2.164). Therefore, some doubts can be raised as to the validity of the asymptotic formulas (2.163) and (2.164) in this case. Fock [1] calculated the propagation constants given by Eq. (2.164) and the exact equation

$$g_1(0, t) = 0, \tag{2.165}$$

expressed via functions of a parabolic cylinder. A sample of the calculations of propagation constants for the hyperbolic profile

$$U(h) = \frac{(h - H_s)^2}{h + h_l}, \tag{2.166}$$

where h_l is a parameter , is presented in Table 2.1

Table 2.1 Comparison of the calculated propagation constants using the asymptotic Eq. (2.164) and the exact formula (2.165).

$H_s + h_l$	Asymptotic t_1	Exact t_1
23.11	–0.085 + j0.466	–0.107 + j0.443
25.24	–0.148 + j0.262	–0.158 + j0.269
48.07	–0.104 + j0.224	–0.113 + j0.227

As observed, the agreement between the asymptotic formulas and the exact solution is satisfactory, especially with regard to the imaginary part of the propagation constants, $\mathrm{Im}(t_n)$. The relation between the attenuation rate of the nth mode, γ_n, and $\mathrm{Im}(t_n)$ is given by

$$\gamma_n = 0.434 \frac{k}{m^2} \,\mathrm{Im}(t_n), \ \mathrm{dB/km}. \tag{2.167}$$

Consider Eq. (2.164) for trapped modes which have a small imaginary component of the propagation constant $t_n$and are in the interval: $U(H_s) \le \mathrm{Re}\, t_n < U(0)$. The respective values of parameter ν are negative. For $\nu < 0$ we define $\ln(\nu) = \ln(-\nu) + j\pi$ and from Eq. (2.152) we obtain

$$S_0 = \int_0^{b_1} \sqrt{U(h) - t}\,dh - j\frac{\pi}{2}\nu \equiv S_1 - j\frac{\pi}{2}\nu. \tag{2.168}$$

Equation (2.164) then takes the form

$$je^{2jS_1} = \chi(\nu) \cong 1\,. \tag{2.169}$$

For large negative ν, $\chi(\nu) \to 1$, and we obtain

$$S_1 \equiv \int_0^{b_1} \sqrt{U(h) - t}dh = \left(m - \frac{1}{4}\right)\pi, \quad m = 1, 2, \ldots \tag{2.170}$$

For either large positive ν or complex ν with positive Re ν, it follows that Re $t < U(H_s)$. In this case the phase S_1 can be defined as

$$S_1 = S_0 + j\frac{\pi}{2}\nu \tag{2.171}$$

and Eq. (2.164) will again be truncated to Eq. (2.170).

A basic conclusion followed from the study of Eq. (2.164) and its truncated form (2.170) (performed by Fock in Ref. [1]) is that the attenuation rate of the field in the evaporation duct is affected not only by the height Z_s and the M-deficit of the evaporation duct but also by a curvature of the M-profile at the level of inversion height Z_s. The attenuation rate γ_m of the mth mode from Eq. (2.170) can be estimated as follows:

$$\gamma_m = 2\pi\left(m - \frac{1}{4}\right)10^{-3}\frac{Z_s M''(Z_s)}{\beta\sqrt{2(M(0) - M(Z_s))}}\Theta = \frac{2\pi}{\lambda_m}10^{-6} Z_s^2 M''(Z_s)\Theta \tag{2.172}$$

where

$$\Theta = \frac{\lambda_m}{\lambda}\operatorname{Im} t_m; \tag{2.173}$$

$$\lambda_m = \frac{\beta\, 10^3 Z_s}{m - 1/4}\sqrt{2(M(0) - M(Z_s))}$$

is a critical wavelength of the mth mode trapped in the evaporation duct formed by a given M-profile, λ is the wavelength of the radiated field. The parameter β is introduced by

$$\beta = \int_0^1 q(z)dz; \quad q(0) = 4; \quad q(1) = 0; \tag{2.174}$$

$$q\left(\frac{z}{Z_s}\right) = 4\frac{M(z) - M(Z_s)}{M(0) - M(Z_s)}. \tag{2.175}$$

For most M-profiles the value of β is close to 1.

References

1 Fock, V.A. *Electromagnetic Diffraction and Propagation Problems*, Pergamon Press, Oxford, 1965.
2 Fock, V.A. Distribution of currents excited by a plane wave on the surface of conductor, *J. Exp. Theor. Phys.*, (Nauka), 1945, 15 (12), 693–702.
3 Leontovitch, M.A. Method of solution to the problems of em wave propagation over the boundary surface, *Izv. Acad. Nauk, Ser. Fiz.*, 1944, 8 (1), 16–22.
4 Leontovitch, M.A., Fock, V.A. Solution to a problem of em wave propagation over the Earth's surface using the method of parabolic equation, *J. Exp.Theor. Phys.*, (Nauka), 1946, 16 (7), 557–573.
5 Fock, V.A. Theory of wave propagation in a non-uniform atmosphere for elevated source, *Izv. Acad. Nauk, Ser. Fiz.*, 1960, 14 (1), 70–94.
6 Krasnushkin, A.E. The expansion over normal waves in a spherically-stratified medium, *Dokl. Acad. Nauk SSSR*, 1969, 185 (6), 1262–1265.
7 Kravtsov, Yu.A., Orlov, Yu.I., Geometric Optics of Non-uniform Media, Nauka, Moscow, 1980, 340 pp.
8 Rotheram, S. Radiowave propagation in the evaporation duct, The Marconi Rev., 1974, 67 (12), 18–40.
9 Booker, H.G., Walkinshaw W. The mode theory of tropospheric refraction and its relation to waveguides and diffraction, in *Meteorological Factors in Radiowave Propagation*, The Royal Society, London, 1946, pp. 80–127.
10 Andrianov, V.A. Diffraction of UHF in a bilinear model of the troposphere over the Earth's surface, *Radiophys. Quantum Electron.*, 1977, 22 (2), 212–222.
11 Wait, J.R. *Electromagnetic Waves in Stratified Media*, Pergamon Press, Oxford, 1962, 372 pp.
12 Hitney, H.V., Pappert, R.A., Hattan, C.P. Evaporation duct influences on beyond-the-horizon high altitude signals, *Radio Sci.*, 1978, 13 (4), 669–675.
13 Chang, H.T. The effect of tropospheric layer structures on long-range VHF radio propagation, *IEEE Trans. Antennas and Propagation*, 1971, 19 (6), 751–756.
14 Wait, J.R., Spies, K.P. Internal guiding of microwave by an elvated tropospheric layer, *Radio Sci.*, 1964, 4 (4), 319–326.
15 Gerks, I.H. Propagation in a super-refractive troposphere with a trapping surface layer, *Radio Sci.*, 1969, 4 (5), 413–417.
16 Bezrodny, V.G., Nicolaenko, A.P. and Sinitsin, V.G. Radio propagation in natural waveguides, *J. Atmos. Terr. Phys.*, 1974, 39 (5), 661–688.
17 Klyatskin, V.I. *Method of Submergence in the Theory of Wave Propagation*, Nauka, Moscow, 1985, 256 pp.
18 Lifshitz, I.M., Gredeskul, S.A. and Pastur, L.A. *Introduction to the Theory of Disorganised Systems*, Nauka, Moscow, 1982, 360 pp.
19 Bass, F.G., Fuks, I.M. *Wave Scattering from a Statistically Rough Surface*, Pergamon press, New York,1979, 424 pp.
20 Rytov, S.M. *Introduction to Statistical Radiophysics, Part 1, Random Processes*, Nauka, Moscow, 1976, 496 pp.
21 Rytov, S.M., Kravtsov, Yu. A. and Tatarski V.I. *Introduction to Statistical Radiophysics, Part 2, Random Fields*, Nauka, Moscow, 1978, 464 pp.
22 Klaytskin, V.I., *Stochastic Equations and Waves in Randomly Non-uniform Media*, Nauka, Moscow, 1980, 336 pp.
23 Isimaru, A. *Wave Propagation and Scattering in Random Media*, Vol.1 and 2, Academic Press, New York, 1978.
24 Tatarski, V.I., Gertsenshtein, M.E. Wave propagation in media with strong fluctuations in refractive index, *J. Exp. Theor. Phys.*, 1963, 44 (2), 676–685.
25 Tatarski, V.I. The EM wave propagation in the media with strong fluctuations in refractive index, *J. Exp. Theor. Phys.*, 1964, 46 (4), 1399–1411.
26 Chernov, L.A. The equations for statistical moments of the field in a random media, *Acoust. J.*, 1969, 15 (4), 594–603.
27 Chernov, L.A. *Waves in Random Media*, Nauka, Moscow, 1975, 172 pp.
28 Shishov, V.I. To the theory of wave propagation in random media, *Radiophys. Quantum Electron.*, 1968, 11 (6), 866–875.
29 Tatarski, V.I. Markov process approximation for light propagation in the medium with random fluctuations in refractive index, *J. Exp. Theor. Phys.*, 1969, 56 (6), 2106–2177.
30 Klyatskin, V.I., Applicability of the Markov random process approximation in the problems of light propagation in the medium with random inhomogeneities, *J. Exp. Theor. Phys.*, 1969, 57 (3), 952–958.

31 Klyatskin, V.I., Tatarski, V.I. Parabolic equation approximation in the problems of wave propagation in the medium with random inhomogeneities, *J. Exp. Theor. Phys.*, 1970, 57 (2), 624–634.

32 Bass, F.G. About the tensor of dielectric permittivity in a medium with random inhomogeneities, *Radiophys. Quantum Electron.*, 1959, 2 (6), 1015–1016.

33 Kaner, E. To the theory of the wave scattering in the medium with random inhomogeneities, *Radiophys. Quantum Electron.*, 1959, 2 (5), 827–829.

34 Pao-Liu Chow, Application of functional space integrals to problems of wave propagation in random media, *J. Math. Phys.*, 1972, 13 (8), 1224–1236.

35 Zavorotny, V.U., Klyatskin, V.I. and Tatarski, V.I. Strong fluctuations of em waves intensity in random media, *J. Exp. Theor. Phys.*, 1977, 73, 481–488.

36 Dashen, R. Path integrals for waves in random media, *J. Math. Phys.*, 1979, 20 (5), 894–918.

37 Gelfand, I.M., Yaglom, A.M. Integration in functional space and its application in quantum mechanics, *Usp. Math. Nauk*, 1956, 11 (1), 77–114.

38 Johnson, G.W., Lapidus, M.L., *The Feynman Integral and Feynman's Operational*, Oxford University Press, Oxford, 2002, 792 pp.

39 Hitney, H.V., Richter, J.H., Pappert, R.A., Anderson, K.D. and Baumgartner, G.B. Tropospheric radio propagation assessment, *Proc. IEEE*, 1985, 73 (2), 265–283.

40 Rotheram, S. Beyond the horizon propagation in the evaporation duct-inclusion of the rough sea, *Marconi Tech. Rep.*, 1974, 74/33, 35 pp.

41 Beckmann, P., Spizzichino, H. *The Scattering of Electromagnetic Waves from Rough Surfaces*, Pergamon and MacMillan, New York, 1963.

42 Tatarskii, V.I. *The Effects of the Turbulent Atmosphere on Wave Propagation*, IPST, Jerusalem, 1971.

43 Baz, A.I., Zeldovich, Y.B., Perelomov, A.M. *Scattering, Reactions and Decays in Nonrelativistic Quantum Mechanics*, Academic Press, New York, 1980.

44 Landau, L.D., Lifshitz, E.M. *Quantum Mechanics (Non-Relativistic Theory)*, Course of Theoretical Physics, Vol. 3, Pergamon Press, Oxford, 2000.

45 Feynman, R.P., Hibbs, A.R. *Quantum Mechanics and Path Integrals*, McGraw-Hill, New York, 1965.

46 Feyzulin, Z.I , Kravtsov, Yu. A. *Radiophys. Quantum Electron.*, 1967, 10, 33–35.

47 Fante, R.L. *Proc. IEEE*, 1975, 63, 1669–1692.

48 Collins, M.D., Evans, R.B., A two-way parabolic equation for acoustic back scattering in the ocean, *J. Acoust. Soc. Am.*, 1992, 91, 1357.

49 Claerbout, J., *Fundamentals of Geophysical Data Processing: With Applications to Petroleum Prospecting*, McGraw Hill, New York, 1976.

50 Borsboom, P.P., Zebic-Le Hyaric, A. RCS predictions using wide angles PE codes, *10th ICAP, 14-17 April 1997, Conference Publication No 436. Proceedings*, pp. 2.191–2.194.

51 Zebic-Le Hyaric, A. Wide-angle nonlocal boundary conditions for the parabolic wave equation, 2001, *IEEE Trans. Antennas Propagation*, 2001, 49 (6), 916–922.

52 Levy, M.F., Borsboom, P.P. Radar cross-section computations using the parabolic equation method, 1996, *Electron. Lett.*, 1996, 32 (13), 1234–1236.

53 Tsuchiya T., T. Anuda T. and Endoh N. An efficient method combining the Douglas operation scheme with the split-step Pade approximation of a higher-order parabolic equation, 2001, *Proc. IEEE Ultrasonics Symp.*, 2001, 683–686.

54 Kuttler, J.R., Dockery, D. Theoretical description of the parabolic approximation/Fourier split-step method of representing electromagnetic propagation in the troposphere, *Radio Sci.*, 1991, 26 (2), 381–393.

55 Barrios, A.E. A terrain parabolic equation model for propagation in the troposphere, *IEEE Trans. Antennas Propagation*, 1994, 42 (1), 90–98.

56 Slingshby, P.I. Modelling tropospheric ducting effects on VHF/UHF propagation, *IEEE Trans. Broadcasting*, 1991, 37 (2), 25–34.

57 Levy, M.F. Parabolic equation modeling of propagation over irregular terrain, *Electron. Lett.*, 1990, 26 (14), 1153–1155.

58 Levy, F.M. Transparent boundary conditions for parabolic equation solutions of radiowave propagation problems, *IEEE Trans. Antennas Propagation*, 1997, 45 (1), 66–72.

59 Janaswamy, R. A curvilinear coordinate-based split-step parabolic equation method for propagation predictions over terrain, *IEEE Trans. Antennas Propagation*, 1998, 46 (7), 1089–1097.

60 Bellis, A., Tappert, F.D. Coupled mode analysis of multiple rough surface scattering, *J. Acoust. Soc. Am.*, 1979, 66 (3), 811– 826.

3
Wave Field Fluctuations in Random Media over a Boundary Interface

In this chapter we study wave propagation over the earth's surface in a line-of-sight region in the presence of a random component of refractive index. In the absence of fluctuation this problem may be regarded as having been solved many decades ago and numerous publications are available. Nonetheless, it may be noted than even in a classical formulation of the problem, i.e. just a point source of vertical or horizontal polarization above a smooth terrain in the absence of super-refraction, the actual solution is not quite simple and can be described in terms of reflection formulas only in a "true" line-of-sight region, at distances not too close to the horizon. Fock [1] demonstrated that the reflection formulas in a form of superposition of direct and reflected waves are valid at distances of the order of a/m before the horizon $\sqrt{2a}\left(\sqrt{z}+\sqrt{z_0}\right)$, i.e. outside the "shade cone". In the case of super-refraction, single reflection formulas are also valid in the range of distances before the horizon, which, in turn, is also modified in the presence of refraction. In particular, in the case of the evaporation duct, there are two separate horizons for the direct and reflected waves [1]. In practical applications [2, 3], signal strength calculations in a line-of-sight region are performed by means of ray theory that can be applied to a very general profile of refractivity.

The problem becomes even more complicated in the presence of random fluctuations in the refractive index. The number of publications on this topic is rather limited and all studies known to the author are concerned with a plane boundary interface [4–6]. The applicability of the results discussed in Ref. [4] is limited to "weak" fluctuations of the scattered field. Strong fluctuations in the scattered field are considered in Ref. [5] using the method of "local perturbations". In Ref. [6] the authors use a perturbation theory for auxiliary functions, treating direct and reflected waves separately. That approach resulted in obtaining an expression for fluctuations in amplitude and phase applicable in the region of the interference minima.

In this chapter we use path integrals to obtain the second- and fourth order moments for the wave field in a random medium above the plane and spherical boundary interface.

Radio Wave. Alexander Kukushkin

ISBN: 3-527-40458-9

3.1
Reflection Formulas for the Wave Field in a Random Medium over an Ideally Reflective Boundary

3.1.1
Ideally Reflective Flat Surface

Let us examine the field in a randomly non-uniform medium above a plane interface. In this case, the Green function $G(\vec{r},\vec{r}_0)$ can be presented as a superposition of the Green function for a point source $G^{+}(\vec{r},\vec{r}_0)$ situated at the point $\vec{r}_0 = \{0, y_0, z_0\}$ and that for the mirror-reflected point source $G^{-}(\vec{r},\vec{r}_0)$ located at $\vec{r}_0 = \{0, y_0, -z_0\}$:

$$G(\vec{r},\vec{r}_0) = G^{+}(\vec{r},\vec{r}_0) - G^{-}(\vec{r},\vec{r}_0)\,. \tag{3.1}$$

The boundary condition at the surface $z = 0$

$$G(\vec{r},\vec{r}_0)\big|_{z=0} = 0\,. \tag{3.2}$$

The boundary conditions (3.2) have the following impact on a continual representation of the Green function: the mirror reflection of the point source requires a mirror reflection of the medium as well, it results in the introduction of the fluctuations in dielectric permittivity to be a function of the modulus of the z-coordinate normal to the surface of separation, $z = 0$, i.e. $\delta\varepsilon(\vec{r}(x)) = \delta\varepsilon(x, y(x), |z(x)|)$. We will define $\vec{\rho}^{\,\pm}(x)$, the trajectories of the waves departing from the sources in the upper and lower half-space. For $G^{\pm}\left(\vec{r},\vec{r}_0\right)$ we have

$$G^{\pm}(\vec{r},\vec{r}_0) = \int D\vec{\rho}^{\,\pm}(x)\exp\left\{jkS_0\left[\vec{\rho}^{\,\pm}(x)\right] + j\frac{k}{2}\int_0^x dx'\,\delta\varepsilon\left(x',\rho^{\pm}\left(x'\right)\right)\right\}, \tag{3.3}$$

We have introduced the notation $\vec{\rho}^{\,\pm}(x) = \left\{y^{\pm}(x), z^{\pm}(x)\right\}$. The action S_0 is given by

$$S_0[\vec{\rho}(x)] = \int_0^x dx'\left[L_0\left(x',\vec{\rho}\left(x'\right)\right) + \frac{1}{2}\delta\varepsilon\left(x',\vec{\rho}\left(x'\right)\right)\right] \tag{3.4}$$

where $L_0(x,\vec{\rho}(x)) = \dfrac{1}{2}\left(\dfrac{d\rho}{dx}\right)^2$. As we can see from Eq. (3.3), the expressions for G^{+} and G^{-} are of identical form but differ in terms of the initial conditions for the trajectories: $\vec{\rho}^{\,+}(x=0) = \{y_0, z_0\}$, $\vec{\rho}^{\,-}(x=0) = \{y_0, -z_0\}$.

Consider a calculation for the second moment of the Green function (3.1). Using Eq. (3.3), we obtain:

$$\left\langle G(\vec{r},\vec{r}_0)G^{*}(\vec{r},\vec{r}_0)\right\rangle = \langle G_{11}\rangle + \langle G_{22}\rangle - \langle G_{12}\rangle - \langle G_{21}\rangle \tag{3.5}$$

where

$$\langle G_{11}\rangle = \int D\vec{\rho}_1^{\,+}(x) \int D\vec{\rho}_2^{\,+}(x) \times \exp\left\{jk\left[S^+(1) - S^+(2)\right] - M_{11}\left[x, \vec{\rho}_1^{\,+}(x), \vec{\rho}_2^{\,+}(x)\right]\right\}, \tag{3.6}$$

$$\langle G_{12}\rangle = \int D\vec{\rho}_1^{\,+}(x) \int D\vec{\rho}_2^{\,-}(x) \times \exp\left\{jk\left[S^+(1) - S^-(2)\right] - M_{11}\left[x, \vec{\rho}_1^{\,+}(x), \vec{\rho}_2^{\,-}(x)\right]\right\}, \tag{3.7}$$

$$M_{11} = \frac{\pi k^2}{4}\int_0^x dx' H\left(\vec{\rho}_1^{\,+}\left(x'\right) - \vec{\rho}_2^{\,+}\left(x'\right)\right), \tag{3.8}$$

$$M_{12} = \frac{\pi k^2}{4}\int_0^x dx' H\left(\vec{\rho}_1^{\,+}\left(x'\right) - \vec{\rho}_2^{\,-}\left(x'\right)\right). \tag{3.9}$$

Here the subscripts 1 and 2 correspond to the first and second sources, respectively; the symbolic notations $S^{\pm}(1)$ and $S^{\pm}(2)$ have the meaning of the actions from the direct source (1) and mirror-reflected source (2). The remaining terms in Eq. (3.5) can be written in a way similar to Eqs. (3.6) and (3.7).

Thus, the coherence function for the field of the point source above the ideally reflective surface is represented by the superposition of correlators between the direct and reflected sources. In the functional space of the trajectories $z(x) = 1/2(z_1(x) + z_2(x))$ and $\varsigma(x) = z_1(x) - z_2(x)$ we can isolate two regions in the presence of a reflective surface.

$$\Omega_1(x) \supset \left\{z^2(x) > \frac{\varsigma^2(x)}{4}\right\}, \quad \Omega_2(x) \supset \left\{z^2(x) < \frac{\varsigma^2(x)}{4}\right\}. \tag{3.10}$$

In the region $\Omega_1(x)$ each term in Eq. (3.5) obeys Eq. (2.35) and the integrals in Eq. (3.5) are similar to those in Eq. (2.68). In the region $\Omega_2(x)$ instead of $H(\vec{\rho}(x))$ we need to use $H(y(x), 2z(x))$. In both regions, $\Omega_1(x)$ and $\Omega_2(x)$, the criteria for applicability of the Markov approximation (2.76) and (2.80) remain the same as for an unbounded medium.

Subsequent calculations are performed here for partially saturated fluctuations [7], engendered by atmospheric turbulence. While the mean-square value of the phase fluctuations is large the following two conditions have to be satisfied.

First we assume the smallness of the mean-square value of the fluctuations in the phase difference at the base to be of the order of the Fresnel zone size:

$$\frac{\pi k^2}{4}\int_0^x dx' H\left(\sqrt{\frac{x'}{\kappa}}\right) = 0.73 C_\varepsilon^2 k^{7/6} x^{11/6} << 1. \tag{3.11}$$

The inequality (3.11) means that even in the presence of phase fluctuations the Fresnel zone remains a characteristic region in the plane $x' = const$ from which the rays arrive in phase. As a consequence, no stochastic multipath occurs and integration in Eq. (3.5) is carried along a single ray-tube limited by a Fresnel zone volume.

The second condition is to require a small fluctuation of the arrival angle compared with the angular size of the Fresnel zone. In path integral representation this requirement is equivalent to

$$\frac{\pi k^2}{8}\int_0^x dx'\vec{\rho}'^2\left(x'\right)\frac{d^2H(\vec{\rho})}{d\vec{\rho}^2} \le \frac{\pi k^2}{8}x^2\int d^2\vec{\kappa}_\perp \Phi_\varepsilon(0,\vec{\kappa}_\perp)\vec{\kappa}_\perp^2 = 0.037C_\varepsilon^2 kx^2 l_0^{-1/3} << 1 \tag{3.12}$$

where l_0 is an internal scale of turbulence. Therefore, when inequalities (3.11) and (3.12) hold, the integration in Eq. (3.5) can be performed along non-perturbed trajectories. For turbulence in the atmosphere and radio frequencies above 10 GHz, the inequalities (3.11) and (3.12) hold at distances $x \le 300$ km.

Non-perturbed trajectories $\vec{\rho}(x)$ represent solutions to Euler equations and are given by

$$\vec{\rho}^{\pm}\left(x'\right) = \pm\vec{\rho}_0 + (\vec{\rho} - \vec{\rho}_0)\frac{x'}{x}. \tag{3.13}$$

Introducing the sum- and difference-coordinates of the corresponding points we obtain

$$M_{11} = \frac{\pi k^2 x}{4}\int_0^1 d\xi H(\vec{\rho}\xi + \vec{\rho}_0(1-\xi)), \tag{3.14}$$

$$M_{12} = \frac{\pi k^2 x}{4}\left\{\begin{array}{l}\int_0^{\xi_1} d\xi H(\varsigma_0 + \xi(2z - \varsigma_0), \gamma\xi + \gamma_0(1-\xi)) + \\ \int_{\xi_1}^1 d\xi H(2z_0(1-\xi) + \xi\varsigma, \gamma\xi + \gamma_0(1-\xi))\end{array}\right\}, \tag{3.15}$$

$$M_{22} = \frac{\pi k^2 x}{4}\left\{\begin{array}{l}\int_0^{\xi_2} d\xi H(\varsigma_0(1-\xi) + \xi\varsigma, \gamma\xi + \gamma_0(1-\xi)) + \\ \int_{\xi_2}^{\xi_1} d\xi H(2(z+z_0)\xi - 2z_0\xi, \gamma\xi + \gamma_0(1-\xi)) + \\ \int_{\xi_1}^1 d\xi H\left(\varsigma\xi + \varsigma_0(1-\xi), \gamma\xi + \gamma_0(1-\xi)\right)\end{array}\right\}, \tag{3.16}$$

where $\{\gamma,\varsigma\} = \vec{\rho}_1 - \vec{\rho}_2$, $\{\gamma_0,\varsigma_0\} = \vec{\rho}'_0 - \vec{\rho}''_0$, $\xi_{1,2} = \frac{z_0 \mp \frac{\varsigma_0}{2}}{z \mp \frac{\varsigma}{2} + z_0 \mp \frac{\varsigma_0}{2}}$ are the distances, expressed in units of $x$along the surface from the source to the point of reflection. The equation for M_{21} is similar to Eq. (3.15) while ξ_1 is replaced with ξ_2.

A similar, but rather simple, equation follows for the intensity of the wave field from the point source $J(x,z,z_0) = 4\pi^2/k^2\left\langle|G(1)|^2\right\rangle$, when $\vec{\rho} = \vec{\rho}_0 = 0$:

$$J(x,z,z_0) = \frac{2}{x^2}[1 - \exp(-M_{12}(x,z,z_0))\cos(\Delta S(x,z,z_0))] \tag{3.17}$$

where $\Delta S(x, z, z_0) = 2k\frac{zz_0}{x}$,

$$M_{12} = M_{21} = \frac{\pi k^2 x}{4}\int_0^1 d\xi H(d_p\xi). \tag{3.18}$$

The M_{12} is a mean-square fluctuation in a phase difference between the direct and the reflected waves, $d_p = zz_0/(z + z_0)$ is the maximum possible separation in the vertical plane between the trajectories of the direct wave and the wave reflected from the surface, in fact d_p is the height at which the direct ray passes the point of mirror-reflection for the reflected wave. In the case of turbulence fluctuations in $\delta\varepsilon$ we obtain

$$M_{12}(x, z, z_0) = 0.869 C_\varepsilon^2 k^2 x d_p^{5/3}. \tag{3.19}$$

The expression (3.19) coincides with the structural function of the phase for the base d_p [8] with accuracy to the numerical coefficient.

As observed from Eq. (3.17) the interference-like structure of the field is kept until the direct and reflected waves are correlated in phase, $M_{12} << 1$. It can be noted that the mean square of the phase fluctuations in either direct or reflected waves may not necessarily be small, $\gamma_S >> 1$. As the correlation between the direct and the reflected waves diminishes, the first term in Eq. (3.17) predominates, yielding a field intensity that is twice as large as that in a free space.

3.1.2 Spherical Surface

The second moment for the field over a spherical surface is determined by Eqs. (3.5) to (3.9) where the Lagrangian is

$$L_0(\vec{\rho}(x)) = \frac{1}{2}\left(\frac{dy}{dx}\right)^2 + \frac{1}{2}\left(\frac{dz}{dx}\right)^2 + \frac{|z(x)|}{a}, \tag{3.20}$$

a is the curvature radius of the spherical surface, z is the height above the surface. In fact, the sphere is replaced by a cylinder with an infinitely long generatrix parallel to the y-axis. The applicability of such an approximation is examined in Ref. [1]. The difference between the given problem and that examined in the previous section is found in the presence of the potential term $|z|/a$ in the Lagrangian (3.20). This term takes into account the spherical boundary surface in a parabolic approximation. The presence of this term leads to the introduction of two segments on the trajectory $z^-(x')$of the reflected wave (departing from the imaginary mirror-reflected source), and these two segments are separated by the reflection point x_0.

3.1.2.1 Trajectory Equations

Following Fock [1], let us introduce the dimensionless coordinates

$$\tilde{x} = \frac{mx}{a}, \quad \tilde{x}_0 = \frac{mx}{a}, \quad y = \frac{kz}{m}, \quad y_0 = \frac{kz}{m} \tag{3.21}$$

and consider the ray in the direction upwards from the source. The phase is given by

$$S^{+} = \tilde{x}t + \frac{2}{3}(y - t)^{3/2} - \frac{2}{3}(y_0 - t)^{3/2}. \tag{3.22}$$

The derivative over t gives the equation for the stationary value of t

$$S'^{+} = \tilde{x} - \sqrt{y - t} + \sqrt{y_0 - t} \tag{3.23}$$

that leads to the solution

$$\sqrt{y_0 - t} = \frac{y - y_0 - \tilde{x}^2}{2x}. \tag{3.24}$$

The equation for the trajectory $y^{+}(\tilde{x}')$ becomes

$$y^{+}(\tilde{x}') = y_0 + \tilde{x}'^2 + \frac{\tilde{x}'}{\tilde{x}}\left(y - y_0 - \tilde{x}^2\right) \tag{3.25}$$

or, in physical coordinates,

$$z^{+}(x') = z_0 + \frac{x'^2}{2a} + \frac{x'}{x}\left(z - z_0 - \frac{x^2}{2a}\right). \tag{3.26}$$

For the ray directed downwards from the source we can separate two regions along $\tilde{x}'$: before and after the reflection point $\tilde{x}_0$. For $\tilde{x}' < \tilde{x}_0$ we obtain

$$S^{-} = \tilde{x}'t + \frac{2}{3}(y_0 - t)^{3/2} - \frac{2}{3}\left(y(\tilde{x}') - t\right)^{3/2} \tag{3.27}$$

and the stationary value of the phase is given by

$$S'^{-} = \tilde{x}' + \sqrt{y\left(\tilde{x}'\right) - t} - \sqrt{y_0 - t} = 0. \tag{3.28}$$

This leads to an equation for the ray trajectory in the region $\tilde{x}' < \tilde{x}_0$

$$y\left(\tilde{x}'\right) = y_0 + \tilde{x}'^2 - 2\tilde{x}'\sqrt{y_0 - t}. \tag{3.29}$$

Assume now that $t = t_0$, the solution to Eq. (3.27) for fixed y, y_0 and $\tilde{x}$. In that case $y(\tilde{x}' = \tilde{x}_0) = 0$, where $\tilde{x}_0 \equiv \tilde{x}_0(t = t_0)$, and

$$y\left(\tilde{x}'\right) = y_0 + \tilde{x}'^2 - \frac{x'}{x_0}\left(x_0^2 + y_0\right) \tag{3.30}$$

becomes the equation for the ray trajectory in the region $x' < x_0$, when $\tilde{x}_0$ is used as a stationary parameter instead of t_0.

The equation for $\tilde{x}' \geq \tilde{x}_0$ can be obtained in a similar way using the boundary condition $y(\tilde{x}' = \tilde{x}) = y$:

$$y\left(\tilde{x}'\right) = \left(\tilde{x}' - \tilde{x}_0\right)^2 + \frac{\tilde{x}' - \tilde{x}_0}{\tilde{x} - \tilde{x}_0}\left(y - (\tilde{x} - \tilde{x}_0)^2\right). \tag{3.31}$$

In physical coordinates, Eqs. (3.30) and (3.31) take the form

$$z\left(x'\right) = z_0 + \frac{x'^2}{2a} - \frac{x'}{x_0}\left(\frac{x_0^2}{2a} + z_0\right), \tag{3.32}$$

$$z^-\left(x' > x_0\right) = \frac{\left(x' - x_0\right)^2}{2a} + \frac{x' - x_0}{x - x_0}\left(z - \frac{\left(x - x_0\right)^2}{2a}\right). \tag{3.33}$$

Now we need to find a reflection point x_0 from the equations for the stationary phase (3.28).

The general case solution is obtained by Fock [1] and is provided in Chapter 2. Let us consider here approximate formulas for small $\delta y = y - y_0$, $|\delta y| << 1$. Introduce $t = t_0 + \delta t$ and expand Eq. (3.28) into series over δy and δt. Leaving only first order terms we obtain

$$\begin{aligned} S^{-\prime} &= x - 2\sqrt{y_0 - t_0} + 2\sqrt{-t_0} + \\ &\delta t\left(\frac{1}{\sqrt{y_0 - t_0}} - \frac{1}{\sqrt{-t_0}}\right) - \delta y\frac{1}{2\sqrt{y_0 - t_0}} = 0. \end{aligned} \tag{3.34}$$

Equating the terms of the same order we obtain

$$\sqrt{-t_0} = \frac{4y_0 - x^2}{4x} = \frac{y_0}{x} - \frac{x}{4}, \tag{3.35}$$

$$\delta t = \frac{\delta y}{2}\frac{\sqrt{-t_0}}{\left(\sqrt{-t} - \sqrt{y_0 - t}\right)} = -\delta y\frac{4y_0 - x^2}{4x^2}. \tag{3.36}$$

The stationary value t_0 corresponds to the case of equal heights of the transmitting and receiving antennas above the surface, $\delta y = 0$, in this case the reflection point x_0 is apparently $x_0 = x/2$. The correction term to t_0 is given by Eq. (3.36) and substitution of δt into Eq. (3.28) will lead to the correction term δx to the distance x_0 to reflection point:

$$\delta x = -\delta y\frac{x}{4y_0 + x^2}. \tag{3.37}$$

Similar approximations can be obtained from the general solution (Chapter 2) when the inequality $\delta y\, x/\rho^3 << 1$ holds, which in turn is equivalent to $\delta y << 1$ since x and ρ^3 are terms of the same order of magnitude. Finally, we have

$$\tilde{x}_0 = \frac{\tilde{x}}{2} - (y - y_0)\frac{\tilde{x}}{4y_0 + \tilde{x}^2} \tag{3.38}$$

and

$$x_0 = \frac{x}{2} - (z - z_0)\frac{x}{4z_0 + \frac{x^2}{2m}}. \tag{3.39}$$

3.1.2.2 Moments of the Field

Under the above conditions of Eqs. (3.11) and (3.12) we can use the equations for unperturbed trajectories of direct and reflected waves (3.26), (3.32), (3.33) in the calculation of the field's moments.

As a less cumbersome case of calculation of the second moment, let us consider the intensity of the field. In the line-of-sight region we obtain an equation similar to Eq. (3.17)

$$J(x,z,z_0)=\frac{2}{x^2}\left[1-\exp\left(-M_{12}^{sph}(x,z,z_0)\right)\cos\left(\Delta S^{sph}(x,z,z_0)\right)\right] \tag{3.40}$$

where

$$M_{12}^{sph}=\frac{\pi k^2 x}{4}\int_0^1 d\xi H\left(d_{sph}\xi\right), \tag{3.41}$$

$$\Delta S^{sph}=k\left[-\frac{x^3}{32a}+\frac{(z-z_0)^2}{8x}+\frac{(z+z_0)x}{4a}-\frac{z^2}{2(x-x_0)}-\frac{z_0^2}{2x_0}\right]. \tag{3.42}$$

The parameter d_{sph} is given by

$$d_{sph}(x)=z_0-\frac{x_0^2}{2a}=z-\frac{(x-x_0)^2}{2a}. \tag{3.43}$$

As observed from Eq. (3.43) and the geometry of the problem, parameter $d_{sph}(x)\to 0$ when the distance x between corresponding points approaches the horizon $x\to x_m=\sqrt{2a}(\sqrt{z}+\sqrt{z_0})$. The phase difference $\Delta S^{sph}\to 0$ and the field intensity $J(x,z,z_0)\to 0$ when $x\to x_m$, in accordance with the laws of geometric optics. The numerical value of Eq. (3.41) is given by Eq. (3.19) where the parameter d_p is replaced by $d_{sph}(x)$.

Let us examine the fluctuations in the field intensity $\left\langle J^2\right\rangle$ above either plane or spherical boundaries with ideal reflection. Using representation (3.1) we obtain

$$\begin{aligned}\left\langle J^2\right\rangle=&\left\langle G_{11}^2\right\rangle+\left\langle G_{22}^2\right\rangle+2\langle G_{11}G_{22}\rangle+\left\langle G_{12}^2\right\rangle+\left\langle G_{21}^2\right\rangle+2\langle G_{12}G_{21}\rangle-\\&2\langle G_{11}G_{12}\rangle-2\langle G_{12}G_{22}\rangle-2\langle G_{21}G_{11}\rangle-2\langle G_{21}G_{22}\rangle.\end{aligned} \tag{3.44}$$

As an example, the fourth-order correlator for a direct wave has the form

$$\begin{aligned}\left\langle G_{11}^2\right\rangle=&\int D\vec{R}'(x)\int D\vec{R}''(x)\int D\vec{\rho}'(x)\int D\vec{\rho}''(x)\times\\&\exp\left\{ik\int_0^x dx'\left[\frac{d\vec{R}'}{dx'}\frac{d\vec{\rho}'}{dx'}+\frac{d\vec{R}''}{dx'}\frac{d\vec{\rho}''}{dx'}\right]\right\}\times\\&\exp\left\{-\frac{\pi k^2}{4}\int_0^x dx'\left[H\left(\vec{\rho}'(x')\right)+H\left(\vec{\rho}''(x')\right)\right]\right\}\times\\&\exp\left\{\begin{aligned}&-\frac{\pi k^2}{2}\int_0^x dx'\int d^2\vec{\kappa}_\perp\,\Phi_\varepsilon(0,\vec{\kappa}_\perp)\exp\left(j\vec{\kappa}_\perp\left(\vec{R}'(x')-\vec{R}''(x')\right)\right)\times\\&\left[\cos\left(\vec{\kappa}_\perp\frac{\vec{\rho}'(x')-\vec{\rho}''(x')}{2}\right)-\cos\left(\vec{\kappa}_\perp\frac{\vec{\rho}'(x')+\vec{\rho}''(x')}{2}\right)\right]\end{aligned}\right\}\end{aligned} \tag{3.45}$$

where $\vec{\rho}'(x) = \vec{\rho}_1^{\,+}(x) - \vec{\rho}_2^{\,+}(x)$, $\vec{\rho}''(x) = \vec{\rho}_3^{\,+}(x) - \vec{\rho}_4^{\,+}(x)$, $\vec{R}'(x) = 1/2\left(\vec{\rho}_1^{\,+}(x) + \vec{\rho}_2^{\,+}(x)\right)$, $\vec{R}''(x) = 1/2(\vec{\rho}_3^{\,+}(x) + \vec{\rho}_4^{\,+}(x))$. The remaining terms in Eq. (3.44) have a form similar to Eq.(3.45).

Under the same conditions defined by inequalities (3.11) and (3.12) we can obtain the mean-square value of the fluctuations in intensity of the wave field $\sigma_J^2 = \langle J^2 \rangle - \langle J \rangle^2$:

$$\sigma_J^2 = \frac{2}{x^2}[1 - \exp(-2M_{12}(x, z, z_0)] \times [1 - \exp(-2M_{12}(x, z, z_0)\cos(2\Delta S(x, z, z_0))] \tag{3.46}$$

and for the scintillation factor $\beta_J^2 = \dfrac{\sigma_J^2}{\langle J \rangle^2}$ respectively

$$\beta_J^2 = \frac{1}{2}\frac{[1-\exp(-2M_{12}(x,z,z_0))][1-\exp(-2M_{12}(x,z,z_0)\cos(2\Delta S(x,z,z_0))]}{[1-\exp(-M_{12}(x,z,z_0)\cos(2\Delta S(x,z,z_0))]^2} \tag{3.47}$$

Here the phase difference ΔS and the mean-square of the fluctuations in a phase difference between the direct and reflected waves M_{12} are determined by the respective formulas for a plane or spherical interface, either Eqs. (3.17) and (3.18) or (3.41) and (3.42).

Given the values of z and z_0 let us introduce the points $x_{\min}$ and $x_{\max}$ at a distance x where the unperturbed field has a maximum and minimum respectively. These points can be derived from the equations $\Delta S(x_{\max}, z, z_0) = \pi(2n + 1)$and $\Delta S(x_{\min}, z, z_0) = 2\pi n$. The value of the scintillation factor β_J^2 at those points we define as $\beta_J^2(x = x_{\max}) \equiv \beta_{J\max}^2$ and $\beta_J^2(x = x_{\min}) \equiv \beta_{J\min}^2$ respectively. As observed from Eq. (3.47), if $M_{12} >> 1$ and, therefore, the direct and reflected waves are completely uncorrelated, the fluctuations in intensity are uniform in space. In this case the scintillation factor $\beta_{J\max}^2 = \beta_{J\min}^2 = 1/2$ at both the maximum and minimum of the unperturbed field. In another limiting case when $M_{12} << 1$, the values of the scintillation factor at the minimum and maximum of the unperturbed field are significantly different. $\beta_{J\max}^2 = 1/2\ M_{12}^2$ and $\beta_{J\min}^2 = 2$; and $\sigma_J^2(x_{\min}) = 8M_{12}^2$ while $\langle J(x_{\min}) \rangle^2 = 4M_{12}^2$. Such behaviour of the field fluctuation is of a general nature. When the only phase fluctuates significantly, which is the case, the amplitude fluctuations $\delta A$of the total field at the minimum $x_{\min}$ are contributed by the fluctuation of the phase difference between the direct and reflected waves $\delta A \sim \Delta\phi_{12} = \delta S^{+}(1) - \delta S^{-}(2)$ and $\left\langle (\delta A)^2 \right\rangle \sim \left\langle \Delta\phi_{12}^2 \right\rangle$; $\left\langle J^2(x_{\min}) \right\rangle \sim \left\langle \Delta\phi_{12}^4 \right\rangle$. With normal distribution of the fluctuations in the phase it follows that $\sigma_J^2(x_{\min}) = 2\left\langle \Delta\phi_{12}^2 \right\rangle = 2\langle J(x_{\min}) \rangle^2$, and consequently $\beta_{J\min}^2 = 2$.

3.2
Fluctuations of the Waves in a Random Non-uniform Medium above a Plane with Impedance Boundary Conditions

Let us introduce the Cartesian coordinate system (x,y,z) and the attenuation factor u by equation $E_z = u \cdot e^{jkx}$.

The slow varying field amplitude u is governed by the equation

$$2jk\frac{\partial u}{\partial x} + \Delta_\perp u + \delta\varepsilon(\vec{r})u = 0 \tag{3.48}$$

with the impedance boundary conditions at the surface separating two media

$$\frac{\partial u}{\partial z} + jqu = 0 \tag{3.49}$$

and initial condition (at the source) $u(x = 0, y, z) = (2\pi/k)\,\delta(\vec{\rho} - \vec{\rho}_0)$, where $\vec{r} = \{x, \vec{\rho}\}$, $\vec{\rho} = \{y, z\}$, $\vec{\rho}_0 = \{y_0, z_0\}$, $\Delta_\perp = \partial^2/\partial y^2 + \partial^2/\partial z^2$, $q = \kappa/\sqrt{\varepsilon_g}$, $k = 2\pi/\lambda$, λ is the wavelength, ε_g is the effective dielectric permittivity of the lower half-space $(z < 0)$, $\delta\varepsilon(\vec{r})$ is the random component of the medium's dielectric permittivity in the upper half-space $(z > 0)$, $\langle\varepsilon\rangle = 0$, and the angle brackets denote averaging over the ensemble of the $\delta\varepsilon(\vec{r})$ realisations.

Following the representation introduced by Malyuzhinetz [9], the boundary problem (3.48), (3.49) reduces to a problem with ideal conditions of reflection via introduction of the Malyuzhinetz transformation:

$$u = \overset{+}{u_0}(x, y, z) + \overset{-}{u_0}(x, y, z) + u_q(x, y, z) \tag{3.50}$$

where

$$u_q(x, y, z) = -2jk \cdot e^{jqz} \int\limits_{\infty \cdot e^{j\pi/4}}^{z} d\varsigma \cdot e^{jk\varsigma}\, \overset{-}{u_0}(x, y, -\varsigma). \tag{3.51}$$

Here $\overset{+}{u_0}$ is the field of the incident wave, and $\overset{-}{u_0}$ is the field of the wave generated by the mirrored source. The $\overset{+}{u_0}$ and $\overset{-}{u_0}$ fields determine the solution of the boundary problem (3.1), (3.2) when $|\varepsilon_g| \to \infty$, i.e., the field of the vertical electrical dipole in the case of a randomly non-uniform medium above an infinitely conducting plane. The last term in Eq. (3.50) takes into consideration the corrective factor for the finite value of ε_g which we can refer to the field of the impedance source.

Let us examine the average intensity $I(\vec{r})$ of the scattered field at the point $\vec{r} = \{x, 0, z\}$, excited by a vertical electric dipole situated at the point $\vec{r}_0 = \{0, 0, z_0\}$:

$$I(\vec{r}) = \left\langle |u(\vec{r})|^2 \right\rangle = I_0(\vec{r}) + \left\langle u_q(\vec{r})\overset{+}{u_0}(\vec{r})^* \right\rangle + \left\langle u_q(\vec{r})\overset{-}{u_0}(\vec{r})^* \right\rangle + c.c. + \left\langle |u_q(\vec{r})|^2 \right\rangle. \tag{3.52}$$

Here, $I_0(\vec{r})$ is the average intensity of the wave field above the ideally reflective surface, $I_0(\vec{r}) = \left\langle \left|\overset{+}{u_0} + \overset{-}{u_0}\right|^2 \right\rangle$. The fields $\overset{\pm}{u_0}$ can be represented in the form of

Feynman trajectory integrals [10]. A trajectory integration is described in Section 3.1 for the boundary problem with ideal reflective conditions, and $I_0(\vec{r})$is given by

$$I_0(\vec{r}) = \frac{1}{x^2}\{1 + \exp[-M_{12}(x, z, z_0)] \cos(\Delta S(x, z, z_0))\} \tag{3.53}$$

where $\Delta S = 2kzz_0/x$ represents the phase difference between the direct and reflected waves. The variance of the fluctuations in phase difference between the direct and reflected waves $M_{12}(x, z, z_0)$ is given by

$$M_{12}(x, z, z_0) = M_{12}(x, d) = \frac{\pi k^2 x}{4} \int_0^1 d \cdot H(\xi d) \tag{3.54}$$

where $H(\rho)$ is a structure function of the fluctuations $\delta\varepsilon$, $d = zz_0/(z + z_0)$ is the height at which the direct beam passes above the surface at the point of mirror reflection. Expression (3.53) and the subsequent formulas of this section have been derived using the approximation of smooth perturbations, i.e. on the validity of the inequalities

$$\frac{\pi k^2}{4} \int_0^x H\left(\frac{\sqrt{x'}}{k}\right) dx' << 1; \tag{3.55}$$

$$\frac{\pi k^2 x^2}{8} \int_0^x \Phi_\varepsilon(0, \kappa_\perp)\kappa_\perp^2 d^2\kappa_\perp^2 << 1; \tag{3.56}$$

where $\Phi_\varepsilon(\kappa)$ is a three-dimensional spectrum of fluctuations in the dielectric permittivity $\delta\varepsilon(\vec{r})$, $\kappa = \{\kappa_x, \kappa_\perp\}$ is the wave vector of the fluctuations, $\kappa_\perp = \{\kappa_y, \kappa_z\}$.

Let us examine the correlation function for the field of the impedance source u_q and the direct wave $\overset{+}{u_0}$. After averaging, trajectory integration and expanding the phase of the derived equation in a series over $\nu = z - \varsigma$, we obtain

$$\left\langle u_q \overset{+}{u_0} * \right\rangle = \frac{2jq}{x^2} \exp[j\Delta S(x, z, z_0) - M_{12}(x, d)] \times \int_C^0 \exp\left\{jk\frac{\nu^2}{2x} - jq\nu\left(1 + \sqrt{\varepsilon_g}\tan(\psi)\right) - k\nu\alpha - k\nu^2\vartheta_s^2(x, d)\right\} d\nu \tag{3.57}$$

where the contour C emanates from infinity along the ray $e^{j5\pi/4}$, $\tan(\psi) = (z + z_0)/x$, ψ is the angle of reflection. The coefficient $\vartheta_s^2(x, d) = C_\varepsilon^2 x d^{-1/3}$ has the meaning of the angular width of the scattered field for a base equal to d, the parameter $\alpha = \vartheta_s^2(x, d)/\gamma$ determines the ratio of the angular spectrum width to the width of the interference lobe $\gamma = 1/kd$.

Let us assume that fluctuations $\delta\varepsilon(\vec{r})$ are caused by turbulence and $d > l_0$ where l_0 is an internal scale of the turbulence, and for the structure function $H(\vec{\rho})$ we will assume $H(\vec{\rho}) = C_\varepsilon^2 \rho^{5/3}$, where C_ε is a structure constant for the fluctuations $\delta\varepsilon$. For coefficients α, $\vartheta_s^2(x, d)$ and $M_{12}(x,d)$ we have

$$\alpha = 1.46kxC_\varepsilon^2\left[\begin{array}{l}2^{5/3}\left(\xi_1\xi_2\right)\left(z^{2/3}\xi_1^{5/3}+z_0^{2/3}\xi_2^{5/3}\right)-\frac{5}{8}\left((2z)^{2/3}\xi_1^{8/3}+(2z_0)^{2/3}\xi_2^{8/3}\right)+\\(2z_0)^{2/3}\xi_2^{5/3}\left(1-2\xi_1^2\right)\end{array}\right] \tag{3.58}$$

$$\vartheta_S^2(x,d) = 1.46kxC_\varepsilon^2\left[\begin{array}{c}\frac{11}{6}2^{5/3}\left(z^{-1/3}\xi_1^{8/3}\xi_2^2+z_0^{-1/3}\xi_2^{2/3}\xi_1^3\right)\left(\xi_1-\frac{6}{11}\right)-\\\frac{5}{3}\left((2)^{2/3}(z^{-1/3}\xi_1^{8/3}\xi_2+z_0^{-1/3}\xi_2^{2/3}\xi_1^3)+(2z_0)^{2/3}\xi_2^{8/3}\right)+\\\frac{5}{24}\left((2z)^{-1/3}\xi_1^{8/3}+(2z_0)^{-1/3}\xi_2^{8/3}(1+\frac{4}{5}\xi_2)\right)\end{array}\right] \tag{3.59}$$

$$M_{12}(x,d) = 0.869C_\varepsilon^2k^2xd^{5/3} \tag{3.60}$$

where the parameters ξ_1 and ξ_2 correspond to the distance of the mirror reflection point x_0 from the source and the receiver respectively:

$$\xi_1 = \frac{x_0}{x} = \frac{z_0}{z+z_0} \quad \text{and} \quad \xi_2 = \frac{x-x_0}{x} = \frac{\zeta}{z+z_0}.$$

When $d < l_0$ the structure function can be approximated by $H(\rho) \propto C_\varepsilon^2\rho^2 l_0^{-1/3}$ and the coefficient $\alpha = 0$, $\vartheta_s^2(x,d) \equiv \sigma_s^2(x)$, where $\sigma_s^2(x)$ is a variance of the fluctuations in the angle of incidence in the vertical plane.

Let us calculate the integral (3.57) for large numerical distances, i.e., when $|qx| >> 1$. Integrating by parts, we obtain

$$\left\langle u_q u_0^+\right\rangle = \frac{1}{x^2}\exp\{-M_{12}(x,d)+j\Delta S(x,z,z_0)\}\times$$

$$\left\{\begin{array}{l}-\frac{2}{(1+jg)\left(1+\sqrt{\varepsilon_g}\tan(\psi)\right)}+\frac{2jg(1+j\beta)}{kx(1+jg)^3\left(1+\sqrt{\varepsilon_g}\tan(\psi)\right)^3}+\\\frac{6\varepsilon_g^2}{(kx)^2}\frac{1}{(1+jg)^5\left(1+\sqrt{\varepsilon_g}\tan(\psi)\right)^5}\end{array}\right\} \tag{3.61}$$

Parameter $\beta = 2kx\vartheta_s^2$ determines the ratio of the width of the angular spectrum of the scattered field ϑ_s^2 to the angular size of the Fresnel zone $\theta_F^2 = 1/kx$. In the light of the validity of inequality (3.56) we have $\beta << 1$. The parameter

$$g = \frac{\alpha}{\left(\frac{1}{\sqrt{\varepsilon_g}}+\tan(\psi)\right)}$$

determines the ratio of the angular spectrum width of the scattered field to the product of the angular width of the interference lobe γ and the effective grazing angle of the reflected wave $1/\sqrt{\varepsilon_g} + \tan(\psi)$.

The correlation function of the impedance-source field u_q with reflected field $\bar{u_0}$ is calculated similarly to Eqs. (3.57)–(3.61). Assume that the following inequalities hold true:

$$\max\{kl_0 >> \left|\sqrt{\varepsilon_g}\right|,\ kl_0 \tan(\psi) >> 1\}. \tag{3.62}$$

In this case the trajectories determining the principal contribution to the trajectory integrals, u_q and $\bar{u_0}$ actually coincide, and for the structure function $H(\rho)$ we can use the quadratic approximation $H(\rho) = C_\varepsilon^2 \rho^2 l_0^{-1/3}$. Then $\langle u_q \bar{u_0}^* \rangle$ can be written as follows

$$\langle u_q \bar{u_0}^* \rangle = \frac{1}{x^2} \left\{ \begin{matrix} -\dfrac{2}{\left(1+\sqrt{\varepsilon_g}\tan(\psi)\right)} + \dfrac{2j\varepsilon_g(1+jg_s)}{kx\left(1+\sqrt{\varepsilon_g}\tan(\psi)\right)^3} + \\ \dfrac{6\varepsilon_g^2}{(kx)^2} \dfrac{(1+jg_s)^2}{\left(1+\sqrt{\varepsilon_g}\tan(\psi)\right)^5} \end{matrix} \right\} \tag{3.63}$$

where $g_s^2 = \sigma_s^2(x)/\theta_F^2$. The parameter $\sigma_s^2(x) = 0.82 C_\varepsilon^2 l_0^{-1/3} x$ is a variance of the fluctuations in the arrival angle of the incident wave.

Let us examine the average intensity of the impedance-source field $\left\langle |u_q|^2 \right\rangle$. Using an approach similar to the calculation of Eq. (3.57) we carry out averaging and continuous integration under the condition of validity of the inequalities (3.55) and (3.56). Expanding the phase of the expression under the integral sign into a series over the distance from the upper limit zand retaining the quadratic terms, we obtain

$$\left\langle |u_q|^2 \right\rangle = \frac{4|q|^2}{x^2} \int_0^{\infty e^{j\pi/4}} du \int_{-2u}^{2u} dv \exp \left\{ \begin{matrix} jk\dfrac{v^2}{2x} - jv\left(\dfrac{ku}{x} - \mathrm{Re}(q) - k\tan(\psi)\right) + \\ 2v\ \mathrm{Im}(q) - kv^2\sigma_s^2 \end{matrix} \right\} dv. \tag{3.64}$$

Let us assume here and below that $\mathrm{Im}\,\varepsilon_g = 4\pi\sigma/\omega = 0$, σ is the conductivity of the lower medium ($z < 0$) and $\omega = 2\pi f$ is a circular frequency of radiation. This condition does not limit the generality of the results since the solution in quadratures for the average intensity (3.52) has already been determined by Eqs. (3.53), (3.57) and (3.64). In the meanwhile such an approximation significantly simplifies the asymptotic expressions for the total field. In the case of radio frequencies above 10 GHz and with the sea surface assumed to be a boundary between the two media, the main contribution to ε_g comes from displacement currents and $\mathrm{Im}\,\varepsilon_g << \mathrm{Re}\,\varepsilon_g$. For large distances $|qx| >> 1$ we obtain

$$\left\langle |u_q|^2 \right\rangle = \frac{1}{x^2}\left\{ -\frac{4}{\left(1+\sqrt{\varepsilon_g}\tan(\psi)\right)^2} + \frac{8\varepsilon_g \sigma_s^2}{\left(1+\sqrt{\varepsilon_g}\tan(\psi)\right)^2} \times \right.$$

$$\left.\left(1+\frac{1}{\left(1+\sqrt{\varepsilon_g}\tan(\psi)\right)^2}\right) - \frac{20\varepsilon_g^2}{(kx)^2}\right\}. \tag{3.65}$$

Collecting all components of Eq. (3.52) we obtain

$$I(x,z,z_0) = \frac{1}{x^2}$$

$$\left\{ \begin{array}{l} 1+|R_0|^2 + 2\exp\left[-M_{12}(x,d)\right]\mathrm{Re}\left[R_s e^{j\Delta S}\right] + 2\exp\left[-M_{12}(x,d)\right] \times \\ \mathrm{Re}\left\{ e^{j\Delta S}\left[\frac{2j\varepsilon_g(1+j\beta)}{kx(1+jg)^3\left(1+\sqrt{\varepsilon_g}\tan(\psi)\right)^3} + \frac{6\varepsilon_g^2}{(kx)^2}\frac{1}{(1+jg)^5\left(1+\sqrt{\varepsilon_g}\tan(\psi)\right)^5}\right]\right\} \\ +\frac{12\varepsilon_g^2}{(kx)^2\left(1+\sqrt{\varepsilon_g}\tan(\psi)\right)^5} + \\ \frac{8\sigma_s^2\varepsilon_g}{\left(1+\sqrt{\varepsilon_g}\tan(\psi)\right)^2}\left(1-\frac{1}{1+\sqrt{\varepsilon_g}\tan(\psi)} + \frac{1}{\left(1+\sqrt{\varepsilon_g}\tan(\psi)\right)^2}\right) - \frac{20\varepsilon_g^2}{(kx)^2} \end{array} \right\}$$

(3.66)

Here $R_0 = \sqrt{\varepsilon_g}\tan(\psi) - 1/\sqrt{\varepsilon_g}\tan(\psi)+1$ is a Fresnel coefficient of reflection in the parabolic approximation for the wave field polarized in a plane of incidence. The coefficient

$$R_s = \frac{\sqrt{\varepsilon_g}\tan(\psi) - 1 + jg\left(\sqrt{\varepsilon_g}\tan(\psi)+1\right)}{\left(\left(\sqrt{\varepsilon_g}\tan(\psi)+1\right)\right)(1+jg)} \tag{3.67}$$

also has the meaning of the reflection coefficient with provision for wave scattering in the media above the surface. When the angle of incidence ψ becomes equal to a Brewster angle $\psi = a\tan(1/\sqrt{\varepsilon_g})$, $R_0 = 0$, and the coefficient R_s is defined by the ratio of the angular width of the scattered field to the width of the interference lobe.

$$R_s = \frac{jg}{(1+jg)} \approx j\frac{\vartheta_s^2(x,d)}{2\gamma}\sqrt{\varepsilon_g}. \tag{3.68}$$

Let us consider some limiting cases. One of the limiting cases is when $\delta\varepsilon = 0$.

We assume in Eq. (3.66) that σ_s, g, $\beta = 0$. In the area of applicability of reflection formulas derived by

$$kx\tan(\psi) \approx kd >> \sqrt{\varepsilon_g} \tag{3.69}$$

we can drop the terms of the order of ε_g/kx and $(\varepsilon_g/kx)^2$. Then

$$I(x,z,z_0) = \frac{1}{x^2}\left[1+|R_0|^2+2R_0\cos(\Delta S)\right]. \tag{3.70}$$

For sliding angles with $\sqrt{\varepsilon_g}\tan(\psi) << 1$ and

$$kd << \sqrt{\varepsilon_g} \tag{3.71}$$

the field intensity (3.66) can be expressed in terms of the familiar attenuation function [9] $w(x) = 2j\varepsilon_g/(kx)$:

$$I(x,z,z_0) = I(x) = \frac{1}{x^2}|w(x)|^2. \tag{3.72}$$

In this limiting case, $R_0 \approx -1$, $\Delta S << \varepsilon_g/(kx) << 1$.

Let us examine Eq. (3.66) in the presence of fluctuations of the refractive index, $\delta\varepsilon(\vec{r}) \neq 0$. The variance of the phase difference between the direct and reflected waves can be written in the form of a ratio of the angular width of the scattered field $\vartheta_s^2(x,d)$ at the base d to the angular width of the interference lobe γ:

$$M_{12}(x,d) = \vartheta_s^2(x,d)/\gamma^2. \tag{3.73}$$

Let us define three characteristic distances on the propagation path: x_s, x_g and x_c from the following relationships:

$$g(x_g,d) = 1, \quad M_{12}(x_c,d) = 1, \quad \sigma_s^2(x_s) = 1. \tag{3.74}$$

Then

$$x_s = k^{-1/2}C_\varepsilon^{-1}l_0^{1/8}, \quad x_g = \left[kd^{2/3}C_\varepsilon^2\sqrt{\varepsilon_g}\right]^{-1}, \quad x_c = \left[k^2d^{5/3}C_\varepsilon^2\right]^{-1}. \tag{3.75}$$

One finds that x_c is the distance where the variance of the phase difference at the base d reaches the order of unity, or the angular spectrum of the scattered field becomes wider than the interference lobe; x_s is the distance where the variance of the fluctuation of the angle of incidence σ_s^2 becomes equal to the square of the angular size of the Fresnel zone θ_F^2; and x_g is the distance from which the angular spectrum of the scattered field becomes wider than the product of the width of the interference lobe γ and the effective grazing angle of the reflected wave $\tan(\psi)/\sqrt{\varepsilon_g}$. Because of the validity of inequality (3.56), the range of distances under consideration is such that $x < x_s$. To analyse the value of the parameters we will define the ratios

$$\frac{x_s}{x_g} = k^{1/2}d^{2/3}C_\varepsilon\sqrt{\varepsilon_g}l_0^{1/6}, \tag{3.76}$$

$$\frac{x_s}{x_c} = k^{3/2}d^{5/3}C_\varepsilon\sqrt{\varepsilon_g}l_0^{1/6}, \tag{3.77}$$

from which it follows that in the range of radio frequencies when $k \approx 1\ \text{cm}^{-1}$ and under conditions of the earth's atmosphere $C_\varepsilon^2 = 10^{-14}\ \text{cm}^{-2/3}$, $|\varepsilon_g| \approx 10^2$, the ratio

$x_s/x_g << 1$. Given the fact that we have used the method of smooth perturbations (inequalities (3.8) and (3.56)), we have always $g << 1$, at the same time $x_c/x_s < 1$, when $d > 10^3$ cm.

In the area of sliding angle propagation when inequality (3.71) is satisfied, the parameter $M_{12}(x,d) < 1$ for all $x < x_s$.

Let us assume that $x_c > x_s$, in this case $M_{12}(x,d) < 1$, $g << \min\left\{1, \sqrt{\varepsilon_g}\tan(\psi)\right\}$, and $R_s \approx R_0$. In the interference region (inequality (3.69)) the field intensity can be approximated as follows:

$$I(x,z,z_0) = \frac{1}{x^2}\left[1 + |R_0|^2 + 2R_0\cos(\Delta S)\exp\left[-M_{12}(x,d)\right]\right]. \tag{3.78}$$

In the case where $\sqrt{\varepsilon_g}\tan(\psi) << 1$, when inequality (3.71) holds, $R_s \approx R_0 \approx -1$. In this case there are two small parameters $M_{12}(x,d)$ and ε_g/kx. Let us introduce yet another characteristic distance x_M, where $M_{12}(x_M,d) = \varepsilon_g/kx_M$:

$$x_M^2 = \frac{\varepsilon_\eta}{k^3 C_\varepsilon^2 d^{5/3}}. \tag{3.79}$$

When inequality (3.62) holds, $x_M < x_s$. Assuming $x < x_M$ the field intensity can be approximated by a composition of two terms: one is an attenuation factor $w(x)$ and the other is a weighted variance of the angle fluctuation of the scattered field:

$$I(x,z,z_0) = \frac{1}{x^2}\left[|w(x)|^2 + 8\varepsilon_g\sigma_s^2(x)\right]. \tag{3.80}$$

In the range of distances $x_s < x < x_M$, the main contribution to the sliding angle propagation mechanism is provided by scattering on the non-uniformity of the refractive index, $\delta\varepsilon(\vec{r})$, and the field intensity in that region is given by

$$I(x,z,z_0) = \frac{1}{x^2}\left[(kd)^2\vartheta_s^2(x,d) + 8\varepsilon_g\sigma_s^2(x)\right]. \tag{3.81}$$

In the regime of non-coherent composition of the direct and reflected waves we have $M_{12}(x,d) >> 1$, and $x_c < x_s$. In this range the wave parameter kd^2/x must be sufficiently large $kd^2/x >> 1$. The interference structure of the wave field in this region is entirely distorted by the large fluctuation of the phase difference between the direct and reflected waves and the intensity of the field is composed of two terms:

$$I(x,z,z_0) = \frac{1}{x^2}\left[|w(x)|^2 + 8\sigma_s^2(x)\right]. \tag{3.82}$$

In conclusion, one can note that the limitation caused by the quadratic approximation of the structure function $H(\rho)$ (inequality (3.63)) is not a fundamental one and serves only to simplify the derivation of the asymptotic expansion for the moments $\langle u_q \bar{u}_0^* \rangle$ and $\left\langle |u_q|^2 \right\rangle$. The analytical solution for the coherence function and the moments of higher order can be obtained similarly to the solution for intensity provided in this section, however, the final expressions are exceedingly cumbersome.

3.3 Comments on Calculation of the LOS Field in the General Situation

From the above study it is apparent that in the general case of the stratified troposphere filled with random fluctuations in the refractive index, the line-of-sight field can be calculated in a similar way, applying the ray theory. This approach can, in principle, be applied to calculation of the field in the tropospheric duct at distances of up to a few hops. In the general case of refractivity, the intensity of the line-of-sight field can be presented in the form

$$J(x,z,z_0) = \sum_{n=1}^{N}\sum_{m=1}^{N} A_n A_m^* \exp\left[jk(S_n - S_m) - M_{nm}\right] \tag{3.83}$$

where A_n, S_n are the amplitude and phase of the nth ray, M_{nm} is a structure function of the phase difference of the phases along the ray's trajectories:

$$M_{nm} = 0.73 C_\varepsilon^2 k^2 \int_0^x ds \left| r_n(s) - r_m(s) \right|^{5/3} \tag{3.84}$$

and z, z_0 are the heights of the receiving and transmitting antennas respectively, x is the distance between them. The phase S_n along the nth ray is given by:

$$S_n = j\frac{k}{2}\int_0^x ds \left[\frac{d^2 r_n}{ds^2} + \varepsilon_m(r_n(s))\right] \tag{3.85}$$

and the trajectories $r_n(s)$ are given by a solution to the Euler equation

$$\frac{d^2 r_n}{ds^2} = \frac{d\varepsilon_m}{dr_n} \tag{3.86}$$

with boundary conditions

$$r_n(0) = z_0,\ r_n(x) = z\,. \tag{3.87}$$

Amplitude A_n accounts for the divergence (or convergence) of the rays

$$A_n = A_n(x) = \left[\frac{D_n(0)}{D_n(x)}\right]^{1/2} \tag{3.88}$$

where $D_n(s)$ is a cross-section of the ray tube at a distance s along the nth ray. For the waves reflected from the sea surface, the amplitude A_n is modulated by a reflection coefficient

$$A_n = A_n(x)\cdot R(\theta_n) \tag{3.89}$$

where θ_n is the angle of incidence of the nth wave in the point of reflection. In most cases, Eqs. (3.83)–(3.87) can be solved by applying computer-based algorithms.

Apparently, the above equations are not applicable with small grazing angles in the case of impedance boundary conditions and should be modified before being applied in the vicinity of caustics.

References

1 Fock, V.A. *Electromagnetic Diffraction and Propagation Problems*, Pergamon Press, Oxford, 1965.

2 Hitney, H.V., Richter, J.H., Pappert, R.A., Anderson, K.D. and Baumgartner, G.B. Tropospheric radio propagation assessment, *Proc. IEEE*, 1985, 73 (2), 265–283.

3 Belobrova, M.V., Ivanov, V.K., Kukushkin, A.V., Levin, M.B. and Fastovsky, J.A. Prediction system on UHF radio propagation conditions over the sea, Institute of Radio Astronomy, Ukrainian Acad. Sci., Preprint No 31, 1989, 39 pp.

4 Bass, F.G., Braude, S.Ya., Kaner, E.A. and Men, A.V. Fluctuations of em waves in a troposphere in the presence of the boundary interface, 1961, *Usp. Sov. Phys. Sci.*, 1961, 73 (1), 89–119.

5 Puzenko, A.A., Chaevsky, E.V. Function of mutual coherence in a problem of small grazing angle wave propagation in random medium over boundary interface, *Radiophys. Quantum Electron.*, 1976, 19 (2), 228–239.

6 Kostenko, N.L., Puzenko, A.A. and Chaevsky, E.V. Correlation of the amplitude and phase of the scattered field in case of propagation through turbulent medium over boundary interface, Preprint of the Institute of Radiophysics and Electronics, Ukrainian Academy of Science, No 153, 1980, 36 pp.

7 Dashen, R. Path Integrals for waves in random media, *J. Math. Phys.*, 1979, 20 (5), 894–918.

8 Tatarskii, V.I. *The Effects of the Turbulent Atmosphere on Wave Propagation*, IPST, Jerusalem, 1971.

9 Feinberg, E.L. *Radio Wave Propagation along the Earth's Surface*, Nauka, Moscow, 1961.

10 Feynman, R.P. and Hibbs, A.R. *Quantum Mechanics and Path Integrals*, McGraw-Hill, New York, 1965.

4
UHF Propagation in an Evaporation Duct

The most important characteristic of the troposphere in terms of radio wave propagation is a refractivity profile averaged over the coordinates tangential to the sea surface, i.e., an M-profile in a "stratified" troposphere:

$$M(z) = 10^6 \left[\frac{\varepsilon(z)-1}{2} + \frac{z}{a}\right] = 10^6 \left[\frac{\varepsilon(z)-1}{2}\right] + 0.157z$$

where z is the height above the sea surface, the gradient 0.157 N-units m^{-1} takes into account the curvature of the earth in the case of normal refraction. According to radio-meteorological and refractometer world-wide data [1–3], the gradient of the modified refractivity $g_m = dM/dz$ is less than the critical gradient $g_c = -0.157$ N-units m^{-1} for more than 50% of the time of observation over the sea surface. In those conditions the surface M-inversion forms an evaporation duct, which significantly affects the process of radio wave propagation by trapping the waves radiated in a selective frequency band.

The simplest characteristic of the ducting properties of the surface based M-inversion is the "critical" wavelength λ_c defined as [4]:

$$\lambda_c = \frac{4}{3} 10^{-3} \beta Z_s \sqrt{2(M(0) - M(Z_s))}, \text{ m} \tag{4.1}$$

and

$$\beta = \int_0^1 \sqrt{q(\varsigma)} d\varsigma.$$

Parameter ς is introduced in Ref. [4] as $\varsigma = z/Z_s$ thus providing a meaningful definition for

$$q\left(\frac{z}{Z_s}\right) = 4\frac{M(z)-M(Z_s)}{M(0)-M(Z_s)}$$

with $q(0) = 4$, $q(1) = 0$.

Frequently observed surface M-inversions of height up to 15 m are capable of ducting radiowaves in a cm band with wavelength $\lambda \le \lambda_c \sim 5$ cm. With increasing wavelength the impact of the evaporation duct is weakened, though the attenuation of the waves with wavelength $\lambda > \lambda_c$ may still be significantly less than in the case

Radio Wave. Alexander Kukushkin

ISBN: 3-527-40458-9

of normal refraction. Fock [4] showed that the attenuation of the field in the shadow region follows the exponential law $\exp(-\gamma x)$, where x is the distance from the horizon, and the magnitude of the attenuation exponent γ depends on the curvature of the M-profile at the minimum point at $z = Z_s$[4]:

$$\gamma \propto 2\pi 10^{-3} \frac{Z_s M''(z=Z_s)}{\beta\sqrt{2\Delta M}} \Theta \tag{4.2}$$

where Θ is a coefficient determined by the imaginary part of the propagation constant t_n.

The above estimates are qualitative in nature. To calculate the field we need to solve the boundary problem (2.26), (2.27).

In a stratified troposphere, calculation of the electromagnetic field beyond the horizon is truncated to the determination of the complex propagation constants E_n and the height-gain functions $\chi_n(z)$ of the normal waves. The total field is then composed of a superposition of the normal waves, which converge in a shadow region. The analytical solution to the problem is known only for a few etalon problems with a limited number of selected M-profiles [4, 5]. In the general case the modal formalism will require a numerical solution of the characteristic equation for propagation constants and, in many cases, numerical integration of the differential equations for the height-gain functions [4–7].

This chapter is arranged as follows: First in Section 4.1, we discuss some results of the propagation measurements and comparison with the prediction based on existing propagation models in order to highlight the current status of the theory and the remaining problems in modelling the propagation phenomena. Then we introduce the perturbation theory for the normal waves and propagation constants in Section 4.2. The perturbation theory provides a means of analytic study of the spectrum of the propagation constants and the height-gain functions with small variations of the M-profile from the etalon profile for which the solution is known.

Section 4.3 is dedicated to the determination of the spectrum of normal waves in an evaporation duct. The solution is limited to the case of a stratified troposphere that is commonly used in existing radio coverage prediction systems.

In Section 4.4 we will study the impact of random fluctuations in refractive index on propagation inside an evaporation duct. The results suggest that the impact may be significant in the frequency range 10 GHz and above.

Section 4.5 deals with the height gain structure of the field inside and outside the evaporation duct for the case of scattering on the turbulent fluctuations in the refractive index. It is shown that scattering on random inhomegeneities of the refractive index leads to a smoother height dependence of the field beyond the horizon compared with duct theory. The results seem to be closer to observations.

4.1
Some Results of Propagation Measurements and Comparison with Theory

Systematic studies of radiowave propagation over the sea surface started in the 1940s, driven by the development of radar, and are still underway due both to the importance of the problem and advances in technology and applications. Several comprehensive programs of radio-meteorological measurements performed in the 1940s and 1950s [8–10] still provide a benchmark reference for further studies due to the exceptionally high quality of the obtained results. Among relatively recent studies we may reference [2], see also several references to reports of the Naval Research Laboratory in Ref. [2].

Here we briefly review the major results obtained in Ref. [10]. The propagation measurements were performed at two frequencies 3 GHz and 10 GHz over a sea path extending to several hundred kilometres and were supported by shipboard meteorological measurements of the temperature, water vapor and air pressure. The restored M-profile was then averaged over several measurements along the propagation path and that averaged profile was approximated by a linear-exponential M-profile (Peceris's model [11]) in order to perform a theoretical calculation of the propagation loss and to estimate the attenuation rate of the signal beyond the horizon. A sample of the comparison of the experimental attenuation rates with Peceris's duct model is presented in Figure 4.1 for 3 and 10 GHz measurements.

As observed from Figure 4.1, the measured attenuation rates of the 3 GHz signal tend persistently to be less than predicted from the linear-exponential model of an evaporation duct. In contrast, the measured attenuation rate significantly exceeded the theoretical estimates. The comparison of the absolute values of the signal strength data with the Peceris model was also performed in Ref. [10]. The conclu-

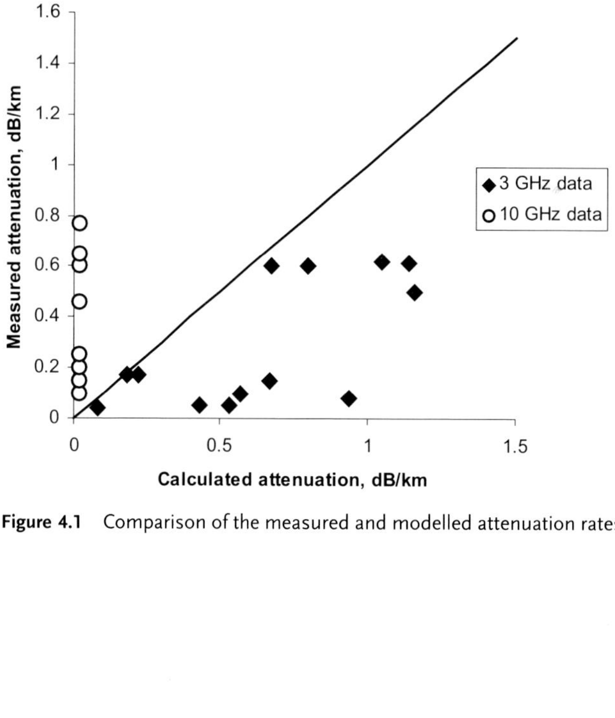

Figure 4.1 Comparison of the measured and modelled attenuation rates at 3 and 10 GHz [10].

sion drawn in 1949 was that there was a significant quantitative discrepancy between the theory and the measurements.

A second revision of the reference data from Ref. [10] was performed in the 80s with close involvement of the author. A new advanced computer-based model of an evaporation duct was implemented to calculate the propagation constants and height gain functions. For comparison of the theory and measurements it was decided to limit to the distances to less than 150 km. The analysis of the measured data led to the conclusion that the evaporation ducts restored at the time of measurement may explain the behavior of the signal in that sub-range, while at larger distances the observed signal levels might be explained by either single-scattering theory in the case of low level signals or the presence of the elevated M-inversion. The last was difficult to analyse since there were no adequate radiosound measurements performed at the time of the radio measurements.

Figure 4.2 shows the result of a comparison of the same measured data as for Figure 4.1 at 3 GHz with a computer-based calculation for bilinear approximation of the averaged M-profile. As observed, the measured attenuation rates are still less than those predicted from the evaporation duct model. Major discrepancies are observed for lower duct heights, in conditions of unstable stratification. One of the measured samples is shown in Figure 4.3, where the restored M-profile reveals an evaporation duct with parameters: $Z_s = 6$ m, and $\Delta M = 2$ N-units. Such discrepancy tends to be persistent in other measurements and cannot just be explained by unavoidable errors in measurements of the meteorological data, see for instance Ref. [12]. We may also refer to the Ref. [5] where similar observations were reported. In principle, the bilinear model tends to underestimate the attenuation rate compared with the more realistic linear-logarithmic model of the evaporation duct. This leads

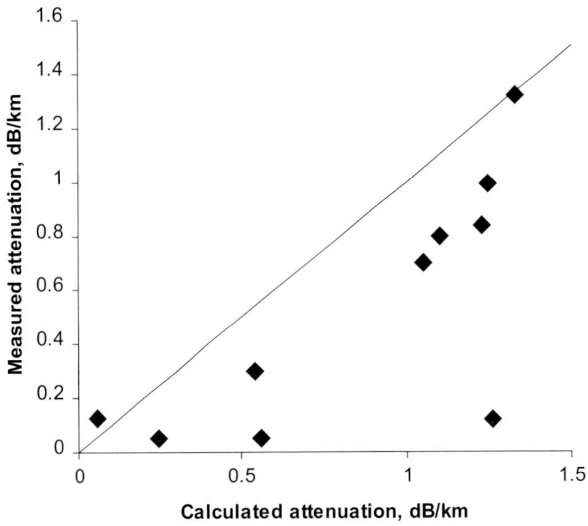

Figure 4.2 Comparison of the measured and modelled attenuation rates at a frequency of 3 GHz. The evaporation duct is modelled by a bilinear M-profile.

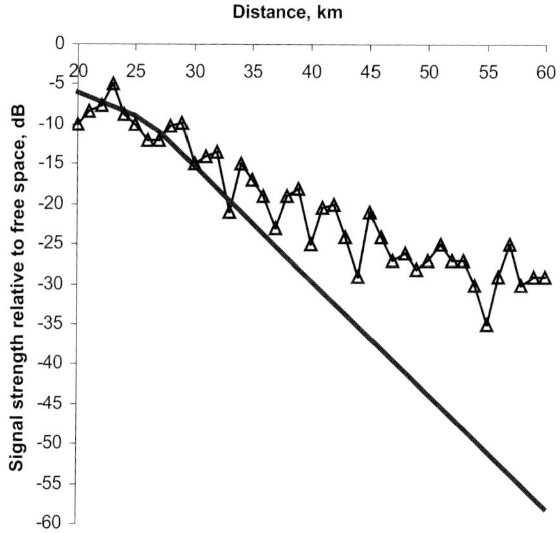

Figure 4.3 Comparison of the measured and predicted signal levels at 3 GHz in the presence of an evaporation duct with height 6 m and M-deficit of 2 N-units.

to the conclusion that the other models of propagation may provide a somewhat intermediate situation when the evaporation duct is not sufficiently strong to support the trapping mechanism by itself, however, negative gradients of the refractivity provide significant enhancement of the propagation beyond the horizon. One of the unusual but possible mechanisms is studied in Chapter 6.

It should be also noted that, in the presence of a strong evaporation duct with heights exceeding 10 m, the waveguide mechanism is more pronounced at a frequency of 3 GHz, and comparison of the evaporation duct model with measurements is rather satisfactory. A typical example is shown in Figure 4.4 for an evapora-

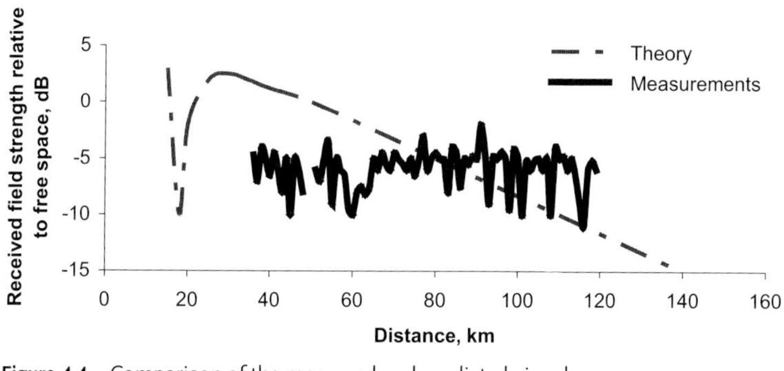

Figure 4.4 Comparison of the measured and predicted signal levels at 3 GHz in the presence of an evaporation duct with height 14 m and M-deficit of 7 N-units.

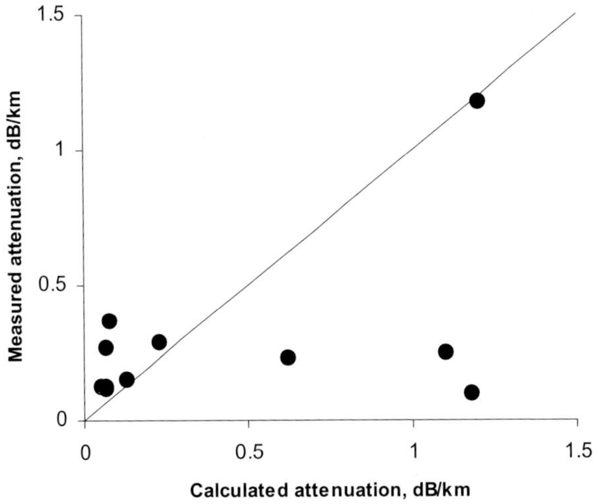

Figure 4.5 Comparison of the measured and modelled attenuation rates at frequency 10 GHz. The evaporation duct is modelled by a bilinear M-profile.

tion duct with height $Z_S = 14$ m, and $\Delta M = 7$ N-units. The heights of the transmitting and receiving antennas are 18 m and 6 m, respectively.

Figure 4.5 shows the comparison of the revised attenuation rates for frequency 10 GHz. As observed, the spread of the data points relative to the bisector is wider than in Figure 4.2 for the 3 GHz data. In most cases the measured attenuation rates are higher than predicted and sometimes the actual propagation mechanism can be

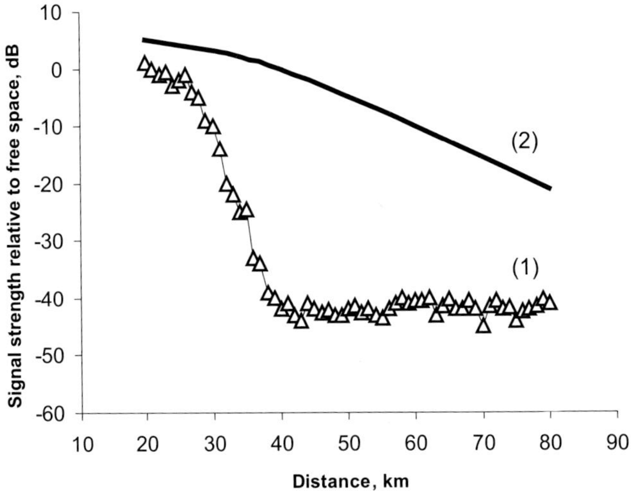

Figure 4.6 Comparison of the measured (1) and predicted (2) signal levels at 10 GHz in the presence of an evaporation duct with height 9 m and M-deficit of 3 N-units.

questioned, for example in Figure 4.6. In general, the evaporation ducts with a height of over 10 m are supposed to provide a waveguide mechanism at 10 GHz with attenuation rate of several hundredth dB km^{-1} (chiefly due to a finite impedance of the sea surface). On the other hand, the attenuation rates measured during the experiment were never less than 0.2 dB km^{-1} at 10 GHz, even in the presence of a very strong and stable with distance evaporation duct.

Figure 4.7 shows the results obtained when the height of the receiving antennas was changed during the experiment. The evaporation duct structure was stable over the distance and the assumption of the uniformity of the duct in a horizontal plane was quite reasonable during the experiment. The evaporation duct with parameters Z_s= 14 m, and ΔM = 7 N-units forms two trapped modes at frequency 10 GHz. The interference between these modes is responsible for the ripples in the theoretical curve for the received signal at a height of 12 m. At a height of 6 m the interference is not pronounced since the second mode has a minimum at this height. A receiver placed above the evaporation duct at a height of 18 m is supposed to receive the em field by penetration through a potential barrier, or, in other words, by a "leak" of the trapped modes. Theoretically predicted signal levels at 18 and 6 m should then differ by 20 dB, as shown in Figure 4.7. On the other hand, the measured data do not reveal any significant difference with the height of the receiving antenna. According to the assessment in Ref. [2] this situation is not unique and is observed in other experiments. One possible explanation of the above phenomena is provided in Section 4.5.

Referring to recent studies described in Refs. [13–20] we may conclude that significant development in modelling the evaporation duct structure in the lower part of the marine boundary layer resulted in the availability of computer-based tools capa-

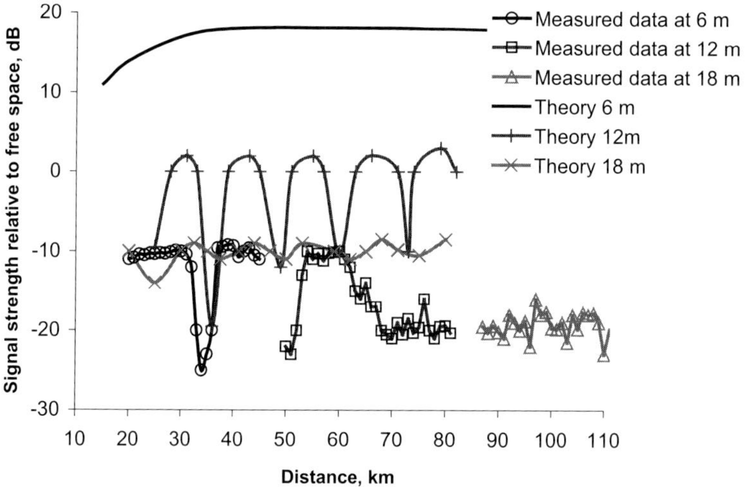

Figure 4.7 Comparison of measured and predicted data at 10 GHz in the presence of an evaporation duct with height 14 m and M-deficit of 7 N-units. The transmitting antenna is mounted at a height of 6 m.

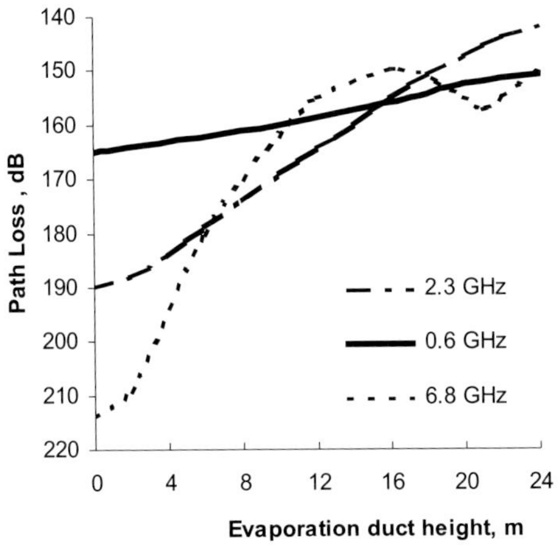

Figure 4.8 Path-loss versus evaporation duct height for the North Sea experiment [16].

ble of reasonably good prediction of the shipboard radar coverage in nearly real time over a wide range of frequency bands from VHF to EHF.

References [16, 17] provide comprehensive analysis of the propagation experiments performed in the Mediterranean and North Sea areas. The statistical assessment of the evaporation duct contribution reported in Ref. [16] also demonstrates unexpectedly high signal levels beyond the horizon in the case of weak ducting conditions at frequencies 0.6 GHz and 3 GHz, where the evaporation ducts with heights less than 10 m are too weak to ensure a single trapped mode. Nonetheless, the received signal tends to be higher than predicted, as shown in Figure 4.8 from Ref. [16]. The same statistics are reported in Ref. [17]. It was suggested in Ref. [16] that the observed high level signals might be caused by propagation mechanisms other than an evaporation duct. One of the conclusions emphasised in Ref. [17], and earlier in Ref. [2], is that the height dependence of the received field is not as significant as may follow from the evaporation duct theory and, basically, high altitude antennas are preferable. In terms of the optimal frequency the results of the measurements suggest the 10–20 GHz band is an optimal band where the evaporation duct makes a strong impact without being counteracted by gaseous absorption at higher frequencies.

A clear demonstration of the advances in radar coverage prediction and range detection enhancement due to an evaporation duct is published in Refs. [2, 12, 14] for 18 GHz radar measurements. Similar results are reported in Ref. [13] for propagation experiments in a range from 3 to 94 GHz. The predicted values of the path loss are depicted as a curve in Figure 4.9. The ripples for the higher duct's height are caused by the presence of several trapped modes and corresponding restructuring of the height gain function of the total trapped field in the waveguide. While overall agreement between prediction and measurements is very good the prediction tends to underestimate the path loss for lower duct heights. It might be noted that the scattering on the rough surface is modelled in a rather conservative way using the Kirchhoff approximation [2, 21], which is applicable to a coherent field component. Nonetheless, an additional

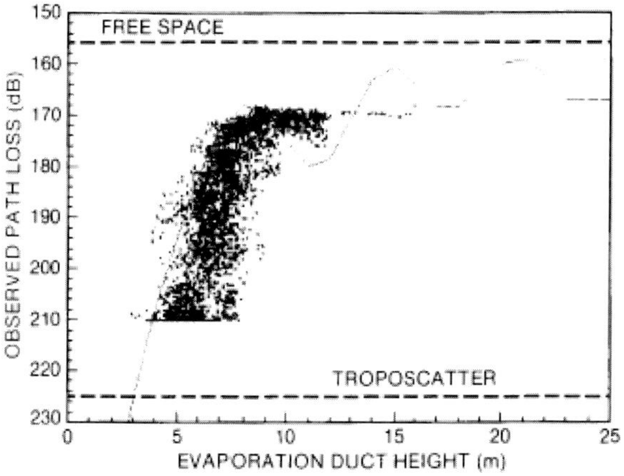

Figure 4.9 Observed path losses for evaporation duct conditions at 17.7 GHz, from Ref. [14].

attenuation factor, like scattering at random fluctuations in the refractive index, might be taken into account for possible improvement in the prediction model. It may also be pointed out that the spread of the data points in Figure 4.9 tends to be more close and symmetrical relative to the prediction curve, which may again suggest the importance of the full treatment of the scattering mechanism. We recall these arguments later in Section 4.3 where the scattering mechanism is described.

Another experiment in the millimetre wave range was reported in Ref. [15] for 94 GHz propagation. The results obtained demonstrated the validity of the evaporation duct mechanism in this frequency band and gave good agreement between the measurements and the theory of the evaporation duct, with some underestimation of predicted path loss with an average value of error of 10 dB. It might be noted that while the prediction model has made use of the Kirchhoff theory of the scattering at a rough sea surface, the model did not take into account scattering on turbulent fluctuations in the refractive index.

4.2
Perturbation Theory for the Spectrum of Normal Waves in a Stratified Troposphere

The presence of stratified inhomogeneities of the refractive index results in a considerable change in the field structure in the region of geometric shadow. A complete solution to the problem of wave propagation in a stratified troposphere, in which the refractive index depends only on the height h above the earth's surface, was obtained by Fock [4]. He showed that for large values of the parameter $m = (ka/2)^{1/3} >> 1$, here k is the wave number and a is the earth's radius, the problem can be simplified by transition from a spherically stratified to a plane-stratified medium. Such a simplification is provided through the use of the parabolic approximation and introduction of the modified permittivity

$$\varepsilon_m(h) = \varepsilon + 2\frac{h}{a}.$$

Even in this simplified formulation an analytical solution can be obtained only for a few standard problems with a limited set of permittivity profiles $\varepsilon_m(h)$. In the general case the analysis of the propagation conditions comes down to a numerical solution for an eigenvalue problem with complex propagation constants. In several cases the function $\varepsilon_m(h)$ differs little from the "etalon" function, as for instance in the case of weak refraction. Therefore, some need arises to construct a perturbation theory for "open" system for which the spectrum of eigenvalues (propagation constants" is complex and eigenfunctions (height factors for the wave field $\chi_n(h)$) grow exponentially with height, $\chi_n(h) \to \infty,\ h \to \infty$. A similar problem arises in quantum mechanics with the study of the decay of quasi-stationary states. Perturbation theory for that case was developed by Zeldovitch [22], although the essential assumption was made on the finiteness of the potential (the analogue of $\varepsilon_m(h)$) at $h \to \infty$. Such a condition is not satisfied in the problem of wave propagation in a stratified troposphere:

$$\varepsilon_m(h) \to 2h/a \text{ as } h \to \infty.$$

Here we describe a generalisation of the perturbation theory [22] to the case of potentials unlimited at $h \to \infty$.

4.2.1 Problem Formulation

The field attenuation factor V for a point source in the spherically stratified medium can be represented by superposition of normal waves

$$V(x, y, y_0) = 2\sqrt{\pi x}e^{j\pi/4} \sum_{n=1}^{\infty} e^{jxt_n} \frac{\chi(y,t_n)\chi(y_0,t_n)\chi^*(0,t_n)}{\frac{\partial}{\partial t_n}\chi(0,t_n)} \tag{4.3}$$

where we introduce dimensionless coordinates $x = mD/a,\ y = kh/m,\ k = k_0\sqrt{\varepsilon(h \to \infty)}$, k_0 is the wave number in a vacuum, and D is the distance along the earth's surface from the source of radiation. The transmitter (point source) is situated at height h_0 and the receiver at height h, respectively. The height functions $\chi(y, t_n)$ are governed by the equation

$$\frac{d^2}{dy^2}\chi(y, t_n) + [U(y) - t_n]\chi(y, t_n) = 0 \tag{4.4}$$

and satisfy the following boundary conditions

$$\chi(0, t_n) = 0, \frac{\delta}{dy}\{\arg(\chi(y, t_n))\}\Big|_{y\to\infty} > 0. \tag{4.5}$$

Equations (4.4) and (4.5) determine a discrete spectrum of propagation constants t_n. We represent the modified refractive index $U(y) = m^2(\varepsilon(y) - 1) + y$, as shown in Figure 4.10, in the form $U(y) = U_0(y) + \delta U(y)$, where $U_0(y)$ is the unperturbed index of refraction for which the solution to the boundary value problem Eqs. (4.4) and (4.5) is known. We shall treat $\delta U(y)$ as the perturbations to the etalon profile $U_0(y)$.

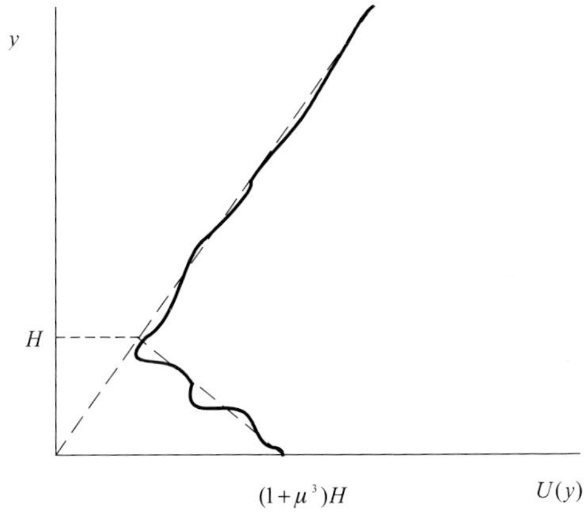

Figure 4.10 Refractivity profile with distortion.

Following the approach in Ref. [22], we introduce the logarithmic derivative

$$z_n(y) = \frac{d\chi(y,t_n)}{dy} \cdot \chi^{-1}(y, t_n) \tag{4.6}$$

through which the height function is expressed as

$$\chi(y, t_n) = \lim_{\sigma\to 0}\chi(\sigma, t_n) \cdot \exp\left(\int^{y} z_n(y')dy'\right). \tag{4.7}$$

For $z_n(y)$ we can obtain the equation

$$z_n'(y) + z_n^2(y) + U_0(y) + \delta U(y) - t_n = 0\,. \tag{4.8}$$

We seek $z_n(y)$ and t_n in the form

$$z_n(y) = \overset{0}{z}_n(y) + \delta z_n(y); \qquad t_n = \overset{0}{t}_n + \delta t_n \tag{4.9}$$

where $\overset{0}{z}_n(y)$ and $\overset{0}{t}_n$ correspond to the height function of the unperturbed Eq. (4.4) and $\delta z_n \sim \delta t_n \sim \delta U$. Substituting Eq. (4.9) into Eq. (4.8), we obtain the equation for correction to the height function

$$\delta z_n' = -2\overset{0}{z}_n(y)\delta z_n(y) + \delta t_n - \delta U(y). \tag{4.10}$$

The solution to Eq. (4.10) has the form

$$\delta z_n(y) = \chi_0^{-2}(y, \overset{0}{t}_n) \int_0^y \chi_0^2(y', \overset{0}{t}_n)\left[\delta t_n - \delta U(y')\right] dy'. \tag{4.11}$$

It is reasonable to assume that perturbations $\delta U(y)$ to a standard profile can be neglected at sufficiently large heights, then the solution to Eq. (4.4) as $y \to \infty$ has the same form as that of the standard problem $\chi_0(y, t_n)$:

$$\delta z_n(y)\Big|_{y\to\infty} \approx \delta\left[\chi_0^{-1}(y, \overset{0}{t_n} + \delta t_n)\frac{d}{dy}\chi_0\left(y, \overset{0}{t_n} + \delta t_n\right)\right] = \frac{\delta t_n}{\chi_o^2(y, \overset{0}{t_n})}\times$$
$$\left[\chi_o(y, \overset{0}{t_n})\frac{\partial^2\chi_o(y, \overset{0}{t_n})}{\partial y\partial t_n} - \frac{\partial\chi_o(y, \overset{0}{t_n})}{\partial y}\frac{\partial\chi_o(y, \overset{0}{t_n})}{\partial t_n}\right]. \tag{4.12}$$

Substituting Eq. (4.12) into Eq. (4.11), we obtain the equation for correction to the propagation constant

$$\delta t_n = N^{-1}\lim\Big|_{y\to\infty}\int_0^y \delta U(y')\chi_n^2(y')dy' \tag{4.13}$$

where $\chi_n(y) \equiv \chi_0(y, \overset{0}{t_n})$.

Equation (4.13) differs from the usual equation of perturbation theory in two aspects: first, there is a $\chi_n^2(y)$ under the integration instead of the square of the absolute value $|\chi_n(y)|^2$ and second, instead of the norm, which does not exist in the present case, there is the finite term N equal to

$$N = \lim\Big|_{y\to\infty}\left\{\int_0^y\left[\chi_n^2(y') - \chi_n(y')\frac{\partial^2\chi_n(y')}{\partial \dot{y}\partial t_n} + \frac{\partial\chi_n(y)}{\partial y'}\frac{\partial\chi_n(y)}{\partial t_n}\right]dy'\right\} = \left(\frac{\partial\chi_n}{\partial y}\frac{\partial\chi_n}{\partial t_n}\right)\Big|_{y=0} \tag{4.14}$$

In the case of the troposphere the variation of the modified refractive index becomes linear at large enough heights and Eq. (4.14) takes the form:

$$N = \lim\Big|_{y\to\infty}\left\{\int_0^y\left[\chi_n^2(y') - w_1^2(\overset{0}{t_n} - y')\right]dy'\right\} - \left[w_1'(\overset{0}{t_n})\right]^2,$$

where $w_1(\overset{0}{t_n} - y)$ is an Airy function [4] which is a solution to the standard problem (4.4) and (4.5) with $U_0(y) = y$.

In the case of normal refraction- $U_0(y) = y, \chi_n(y) = w_1(\overset{0}{t_n} - y)$ for all $0 < y < \infty$. The parameter a shall be understood as equivalent to the earth's radius. The propagation constants in this case are determined from the solution to a transcendent equation $w_1(\overset{0}{t_n}) = 0$ and for large numbers n they have an asymptotic form $\overset{0}{t_n} = \tau_n e^{j\frac{\pi}{3}}, \tau_n = [3/2(n - 1/4)\pi]^{2/3}$.

Equation (4.14) is then reduced to

$$N = \lim\Big|_{y\to\infty}\left\{\int_0^y w_1^2(\overset{0}{t_n} - y')dy' + (\overset{0}{t_n} - y)w_1^2(\overset{0}{t_n} - y) - \left[w_1'(\overset{0}{t_n} - y)\right]^2\right\}. \tag{4.15}$$

Calculating the integral appearing in Eq. (4.15),

$$\int w_1^2(-x)dx = xw_1^2(-x) + \left[w_1'(-x)\right]^2 \tag{4.16}$$

and using the boundary condition (4.5), we obtain $N = -\left[w_1'(\overset{0}{t_n})\right]^2$ and

$$\delta t_n = -\left[w_1'(\overset{0}{t_n})\right]^{-2}\int_0^\infty \delta U(y)w_1^2(\overset{0}{t_n} - y)dy. \tag{4.17}$$

Equation (4.17) allows one to find explicitly the correction terms δt_n to the complex propagation constants $\overset{0}{t_n}$ of the normal wave for arbitrary distortion $\delta U(y)$ of the linear profile of the refractive index. Of course the limitation is that the correction terms prove to be small, i.e. $|\delta t_n| << \left|\overset{0}{t_n} - \overset{0}{t}_{n-1}\right|$. It should be noted that in the case of large perturbation $\delta U(y)$ a new branch of the spectrum of the normal modes could arise, that case is outside the scope of the problem considered here.

4.2.2
Linear Distortion

Let us consider a linear distortion

$$\delta U(y) = \begin{cases} (1+\mu^3)(H-y), & y \le H \\ 0, & y > H \end{cases} \tag{4.18}$$

where $1+\mu^3$ is the gradient of the modified refractive index in the height range $0 < y \le H$ as shown in Figure 4.10. The equation for the spectrum t_n of the normal modes was derived by Fock [4], its solution cannot be obtained analytically and requires application of numerical methods. Application of perturbation theory in form (4.17) allows investigation of the t_n-spectrum in a considerably simple way.

Let the height of distortion be small, assuming that $H << \tau_n$ and $\tau_n >> 1$. Then the phase of the argument of the Airy function lies within the sector $(\pi/3, 2\pi/3)$ and the asymptotic form of $w_1(\overset{0}{t_n} - y)$ is as follows:

$$w_1(\overset{0}{t_n} - y) \sim (\overset{0}{t_n} - y)^{-1/4}\left\{ e^{\frac{2}{3}(\overset{0}{t_n}-y)^{3/2}} + je^{-\frac{2}{3}(\overset{0}{t_n}-y)^{3/2}} \right\}. \tag{4.19}$$

Substitute Eqs. (4.18) and (4.19) into Eq. (4.17) and expand the exponentials in Eq. (4.19) in a series in powers of $y/\overset{0}{t_n}$. Retaining the linear terms, which is valid when inequality $H^2 << \sqrt{\tau_n}$ holds, we obtain

$$\delta t_n = j(1+\mu^3)\left(\overset{0}{t_n}\right)^{-1/2}\left[w_1'(\overset{0}{t_n})\right]^{-2}\left\{ H^2 + \left(4\overset{0}{t_n}\right)^{-1}\left[2 - e^{-2H\sqrt{\overset{0}{t_n}}} + e^{2H\sqrt{\overset{0}{t_n}}}\right]\right\}. \tag{4.20}$$

When the stronger inequality $H^2\tau_n << 1$ applies we can obtain from Eq. (4.20)

$$\delta t_n = -(1+\mu^3)\left\{\frac{H^4}{12} + \frac{H^5\overset{0}{t_n}}{90}\right\}. \tag{4.21}$$

As observed from Eq. (4.21), increase in the height H or depth $(1+\mu^3)H$ of inversion in the refractive index leads to a decrease in the real and imaginary components of the propagation constant t_n, that corresponds to an increase in the phase velocity of the normal wave and a decrease in its attenuation.

The applicability of the perturbation theory is limited by a small variation to the propagation constant

$$|\delta t_n| << \left|\overset{0}{t_n} - \overset{0}{t}_{n-1}\right| \approx \frac{\pi}{\sqrt{\tau_n}}. \tag{4.22}$$

Inequality (4.22) in conjunction with $H^2\tau_n << 1$ yields the limits of applicability of Eq. (4.21):

$$H << \min\left[\tau_n^{-1/2}, \left(\frac{12}{(1+\mu^3)}\right)^{1/4}\tau_n^{-1/8}\right] \tag{4.23}$$

and, for Eq. (4.20)

$$H << \frac{1}{\sqrt{\tau_n}}\ln\left(\frac{16\tau_n^{3/2}}{1+\mu^3}\right). \tag{4.24}$$

Let us consider another limiting case, when the height of distortion is large $H >> \tau_n$. We represent the integral in Eq. (4.17) as the sum of integrals over the arc of radius H in the sector of angles $(0, \pi/3)$ and over the ray $\arg y = \pi/3$:

$$\int_0^H (H-y)w_1^2(\overset{0}{t_n}-y)dy = I_1 + I_2\,,$$

$$I_1 = \int_0^{He^{j\pi/3}} (H-y)w_1^2(\overset{0}{t_n}-y)dy\,,$$

$$I_2 = jH\int_0^{\pi/3} (1-e^{j\varphi})w_1^2(\overset{0}{t_n}-He^{j\varphi})d\varphi\,.$$

In the integral I_1 we make a change in the variable $y = \tau e^{j\pi/3}$ and use the property of the Airy function $w_1(xe^{j\pi/3}) = 2e^{j\pi/6}v(-x)$:

$$I_1 = 4e^{j2\pi/3}\int_0^H (H-\tau e^{j\pi/3})v^2(\tau-\tau_n)dy\,. \tag{4.25}$$

The upper limit of integration in Eq. (4.25) can be extended to infinity since the Airy function $v(\tau-\tau_n)$ decreases exponentially for $\tau > \tau_n$. Then, using Eq. (4.16) we obtain

$$I_1 \approx 4e^{j2\pi/3}H\int_0^\infty v^2(\tau-\tau_n)dy = -4e^{j2\pi/3}H\sqrt{\tau_n}\,. \tag{4.26}$$

To estimate integral I_2 we use the asymptotic behavior of the function $w_1(\overset{0}{t_n}-y)$ for large negative arguments

$$I_2 = -H^2\int_0^{\pi/3} (1-e^{j\varphi})(He^{j\varphi}-\overset{0}{t_n})^{-1/2}e^{j\Psi(\varphi)}d\varphi \tag{4.27}$$

where $\Psi(\varphi) = 4/3H^{3/2}e^{j3\phi/2} - 2\sqrt{H}e^{j\phi/2}\overset{0}{t_n}$. The main contribution to the integral (4.27) comes from the region of small φ where the only linear terms can be retained in the expansion of the integrand. Then we obtain

$$I_2 \cong \frac{1}{4H^{3/2}}\exp\left(j\frac{4}{3}H^{3/2} - 2j\sqrt{H}\overset{0}{t_n} + j\frac{\pi}{2}\right). \tag{4.28}$$

Substituting Eqs. (4.28) and (4.26) into Eq. (4.17) we obtain

$$\delta t_n = (1+\mu^3)H\left[1 - \frac{j}{16\tau_n H^{5/2}} \exp\left(j\frac{4}{3}H^{3/2} - 2j\sqrt{H}\overset{0}{t_n}\right)\right]. \tag{4.29}$$

As observed from Eqs. (4.29) and (4.21), the correction factor δt_n grows exponentially with number n since the imaginary part of the propagation constant is $\operatorname{Im} t_n \sim n^{3/2}$, and hence the inequality (4.22) ceases to be satisfied, starting with certain n. It can be shown, however, that this result is a consequence of the "non-physical" nature of the broken linear profile. For smooth functions $U(y)$ that actually describe realistic behavior of the refraction index, the δt_n-dependence on a wavenumber n has a fundamentally different character (see, for instance, Eq. (4.40) below).

In fact, performing integration by parts in Eq. (4.17), we can write

$$\delta t_n = -N^{-1} \int_0^\infty \frac{d\delta U(y)}{dy} F_n(y)dy \tag{4.30}$$

where $F_n(y) = \int_0^y w_1^2(\overset{0}{t_n} - y')dy'$ behaves as $1/2\overset{0}{t_n} \exp(2\overset{0}{t_n}\sqrt{y})$ for large numbers n: $\sqrt{\tau_n}y >> 1$. If $\delta U(y)$ is a function continuous along y altogether with its derivatives and which decreases sufficiently at infinity, then integration in Eq. (4.30) will result in values of δt_n which decrease for large n.

In the case where $\delta U(y)$ is abruptly reduced to zero at $y = H$ (a discontinuity of the first kind), a singularity of a delta-function type appears in the integrand in Eq. (4.30) at this point, and at large t_n it make the main contribution to the integral:

$$\delta t_n \sim -N^{-1} F_n(y=H) \sim \exp\left(2H\sqrt{\overset{0}{t_n}}\right) \frac{1}{2\left(\overset{0}{t_n}\right)^{3/2}}. \tag{4.31}$$

This result can be generalized to the case of discontinuity of the kth derivative:

$$\delta t_n \sim \exp\left(2H\sqrt{\overset{0}{t_n}}\right) \frac{1}{2\left(\overset{0}{t_n}\right)^{3/2}} \frac{1}{2^k\left(\overset{0}{t_n}\right)^{k/2}}. \tag{4.32}$$

4.2.3
Smooth Distortion

In the atmosphere the averaged dielectric permittivity $\varepsilon(h)$ is a sufficiently smooth function that normally decreases with height. Let us consider a perturbation $\delta U(y)$ described by a continuous function, which permits analytical expansion into the region of complex y. To calculate the correction term it is convenient to use, instead of the function $\delta U(y)$, its Laplace transformation: $F(s) = \int_0^\infty dy e^{-sy} \delta U(y)$:

$$\delta t_n = N^{-1} \int_{\sigma - j\infty}^{\sigma + j\infty} ds F(s) \int_0^\infty e^{sy} w_1^2(\overset{0}{t_n} - y)dy \tag{4.33}$$

Let the function $\delta U(y)$ be such that $F(s)$ has no singularities in the right half plane s or on the imaginary axis. Then the integration contour over ds can be shifted into the left half plane, while the contour in the integral over dy can be deformed into a ray with $\arg(y) = \pi/3$. Substituting the asymptotic form (4.19) into Eq. (4.33) we obtain

$$\int_0^\infty w_1^2(t_n^0 - y)e^{sy}dy = 4e^{j2\pi/3}\int_0^\infty v^2(y-\tau_n)\exp(sye^{j\pi/3})dy + \lim|_{R\to\infty}\left[R^{-3/2}\exp\left(sR+4\frac{\tau_n}{3}\sqrt{R}\right)\right]. \tag{4.34}$$

In line with the above assumptions, $\mathrm{Re}\, s < 0$, so the second term in Eq. (4.32) vanishes. Using an inverse Laplace transformation, we obtain

$$\delta t_n = 4e^{j2\pi/3}N^{-1}\int_0^\infty \delta U(ye^{j\pi/3})v^2(y-\tau_n)dy. \tag{4.35}$$

It should be noted that Eq. (4.35) could be used when the convergence sector of the function $\delta U(y)$ is wider than $\pi/3$.

Let us consider the application of Eq. (4.35) to the example of perturbation given by $\delta U(y) = U_0\exp(-y/H)$. We set $\tau_n >> M1$ and $H << 1$. Substituting the asymptotic form $v(y-\tau_n)$, we obtain

$$\delta t_n = 2U_0\frac{e^{j\pi/3}}{N}\int_0^y dy\exp\left(\frac{-ye^{j\pi/3}}{H}\right)\left[1+\sin\left(\frac{4}{3}(\tau_n - y)^{3/2}\right)\right]. \tag{4.36}$$

When inequality $H^2 << \sqrt{\tau_n}$ is satisfied Eq. (4.36) can be reduced to

$$\delta t_n = 2U_0\frac{e^{j\pi/3}}{N\sqrt{\tau_n}}H\left[1-\frac{1}{1-4\tau_n H^2 e^{-j2\pi/3}}\right]. \tag{4.37}$$

When $H^2\tau_n << 1$ Eq. (4.37) converges to a simple form

$$\delta t_n = -U_0\left[2H^3 + 8\tau_n H^5 e^{j\pi/3}\right]. \tag{4.38}$$

In the opposite case, $H^2\tau_n >> 1$, we obtain

$$\delta t_n \cong \frac{U_0 H}{2\tau_n}e^{-j\pi/3}. \tag{4.39}$$

4.2.4
Height Function

The correction to the height function $\chi_0(y, t_n^0)$ for known δt_n is determined by the integral (4.11), which allows one to obtain simple equations for $\delta z_n(y)$ in various limiting cases.

For example, let us consider the case when the antennas are placed sufficiently low above the earth's surface, i.e. $y \sim y_0$ and $y^2\tau_n << 1$, while the perturbation $\delta U(y)$ is given by the function (4.18). Expanding $\chi_0(y, t_n^0)$in a series in powers of y, we obtain

$$\chi(y, t_n) \approx \chi_0(y, \overset{0}{t_n})\exp\left[\frac{\delta t_n y^2}{6} - (1+\mu^3)\frac{y^3}{36}\right]. \qquad (4.40)$$

As observed from Eq. (4.40), the new height function $\chi(y, t_n)$ is localized near the surface to a greater degree that the unperturbed one $\chi_0(y, \overset{0}{t_n})$.

4.2.5
Linear-Logarithmic Profile at Heights Close to the Sea Surface

Consider another specific kind of perturbation to the refractivity profile, which can be applied to the estimation of the impact from large gradients of humidity in the immediate vicinity of the sea surface. Refractometer measurements or measurements of the meteorological parameters are difficult to perform at these heights because of the roughness of the sea surface. On the other hand, this layer can be characterised by extremely rapid changes in refractivity due to the gradients of humidity. As follows from hydrodynamic theory of the evaporation duct [5, 23], the modified refractivity profile in the layer close to the sea surface can be modelled by a linear-logarithmic function:

$$U(y) = y - Y_s \ln\left(\frac{y}{y_r}\right) \qquad (4.41)$$

where $Y_s = kZ_s/m$ and Z_s is the evaporation duct's height, $y_r = kz_r/m$, and z_r is the height above the sea surface below which it is not feasible to make the measurements. Normally z_r is associated with a sea roughness parameter. Now we consider the impact from the gradient of humidity in a layer below y_r as a perturbation δU to a linear profile:

$$\delta U = \begin{cases} -Y_s \ln\left(\frac{y}{y_r}\right), & y \le y_r \\ 0, & y > y_r. \end{cases} \qquad (4.42)$$

In fact, we may approximate the initial profile (4.41) with a linear segment in the interval $0 < y < y_r$ as discussed later in Section 4.3, and regard the combined profile as the "true" profile while considering the logarithmic part as the perturbation.

With the perturbation in form (4.42) and if $y_r^2 \tau_n << 1$ we obtain the correction term to the propagation constant of the first mode

$$\delta t_n = -\left(\frac{Y_s y_r^3}{9} + \frac{y_r^5 \overset{0}{t_n}}{75}\right). \qquad (4.43)$$

We may now evaluate the reference height z_r below which the exact shape of the profile does not make a difference to the calculation of the field attenuation beyond the horizon. The impact perturbation is negligible if the variation in attenuation of the first mode due to perturbation (4.42) is small, i.e.

$$\frac{kx}{2m^2}\,\mathrm{Im}\,t_n \approx 6\cdot 10^{-6}\frac{k^6 z_r^5}{m^7}x << 1. \tag{4.44}$$

In the range of distances $x \sim 100$ km for frequencies of the order of 10 GHz , we may see that the above inequality is satisfied with $z_r \le 2$ m. Therefore, we may state that the refractivity measurements can be performed from heights above $z_r = 2$ m and the M-profile close to the surface at heights below z_r can be approximated by a linear function or any other function approaching a finite value at $z = 0$.

4.3
Spectrum of Normal Waves in an Evaporation Duct

As shown in Chapter 2, the attenuation function of the wave field produced by a point source in a shadow region is determined by the sum of residues in the poles of the S-matrix. The pole's position in a plane of complex $t = m^2/k^2 E$ is given by the roots of the equation

$$\chi^+(0,t) = 0. \tag{4.45}$$

The function $\chi^+(h,t)$ is a solution to Eq. (2.36) representing the wave outgoing to infinity, $h \to \infty$. The shadow region is defined as the region of the distances $\xi > \xi_h$, where ξ_h is a horizon of the wave reflected once from the earth's surface.

In the case of linear approximation

$$U(h) = \begin{cases} \mu_1^3(H_s - h), & h \le H_s \\ \mu_2^3(h - H_s), & h > H_s. \end{cases} \tag{4.46}$$

Eq. (4.45) can be truncated to the following

$$1 - R_s(t)R_g(t) = 0 \tag{4.47}$$

where

$$R_s = -\frac{\mu_1 w_1(x_2)w_1'(x_1) + \mu_2 w_1'(x_2)w_1(x_1)}{\mu_1 w_1(x_2)w_2'(x_1) + \mu_2 w_1'(x_2)w_2(x_1)}, \tag{4.48}$$

$$R_g = -\frac{w_1(x_0)}{w_2(x_0)},$$

$$x_1 = \frac{t}{\mu_1^2}, \quad x_2 = \frac{t}{\mu_2^2}, \quad x_0 = \frac{\left(t - \mu_1^3 H_s\right)}{\mu_1^2}.$$

Here, R_s and R_g are reflection coefficients of the waves from the boundaries $h = H_s$ and $h = 0$, respectively. It should be noted that we assume an ideal reflection at the boundary $h = 0$, since the boundary impedance q is assumed to be infinite, $|q| \to \infty$.

Note that here and throughout the book we use a "shifted" M- and $\varepsilon_m(z)$ profile to define a non-dimensional profile $U(h)$. In particular, the "shifted" $\hat{\varepsilon}_m(z)$ profile is given by $\hat{\varepsilon}_m(z) = \varepsilon_m(z) - \min\{\varepsilon_m(0), \varepsilon_m(Z_s)\}$. The non-dimensional U-profile is then determined as

$$U(h) = m^2[\hat{\varepsilon}_m(y) - 1] = m^2[\varepsilon(y) - 1 - \min[\varepsilon(0), \varepsilon(H_s)]] + y.$$

The difference between the conventional U-profile and the shifted one is an additional phase shift in the propagation constant that can be accounted for as a phase correction in the definition of the envelope W_1 given by Eq. (2.16).

Consider positive values of t. With $\mathrm{Re}(t) > 0$, in accordance with the general principle of quantum mechanics, the movement of a particle in a potential (4.36) becomes finite, i.e. the wave is reflected from the potential barrier. With $t >> \mu_1^2, \mu_1^2$,

$$R_s \cong -1 + 2jC(t) \tag{4.49}$$

where

$$C = -\frac{\mu_1 w_1(x_2)v'(x_1) + \mu_2 w_1'(x_2)v(x_2)}{\mu_1 w_1'(x_1)w_2(x_2) + \mu_2 w_2'(x_2)w_1(x_1)} \tag{4.50}$$

and Eq. (4.47) is equivalent to

$$v(x_0) - Cw_1(x_0) = 0. \tag{4.51}$$

We can use perturbation theory to solve the above equation for small values of $C(t)$ which correspond to large positive values of the parameter t. Let us assume first that $C = 0$. In this case the roots of the truncated Eq. (4.51) are defined by a series of positive real numbers:

$$\overset{0}{t_n} = \mu_1^3 H_s - \mu_1^2 \xi_n \tag{4.52}$$

where $\xi_1 = 2.338$, $\xi_2 = 4.088$, $\xi_3 = 5.521$; with $n >> 1$, $\xi_n = \left[\frac{3}{2}\pi\left(n - \frac{1}{4}\right)\right]^{2/3}$. The first correction term δt_n to the value of $\overset{0}{t_n}$ takes the form

$$\delta t_n = -\frac{\mu_1^2 C\left(\overset{0}{t_n}\right)}{\left[v'(-\xi_n)\right]^2} \approx \frac{\mu_1^2}{\sqrt{\xi_n}}\left\{\frac{\mu_1^3 + \mu_2^3}{16\left(\overset{0}{t_n}\right)^{3/2}} \exp\left[-\frac{4}{3\mu_1^3}\left(\overset{0}{t_n}\right)^{3/2}\right] + j\gamma_n\right\} \tag{4.53}$$

and

$$\gamma_n = \frac{\mu_1^2}{4\sqrt{\xi_n}} \exp\left\{-\frac{4}{3}\frac{\mu_1^3 + \mu_2^3}{\mu_1^3 \mu_2^3}\left(\overset{0}{t_n}\right)^{3/2}\right\}. \tag{4.54}$$

Parameter γ_n has the meaning of the square of the module of the coefficient of penetration through the potential barrier.

Equation (4.51) has a finite number N of roots with $\mathrm{Re}\, t_n > 0$, which determines the number of the trapped modes in the evaporation duct. This number can be estimated as

$$N = \text{entier}\left[\frac{2}{3\pi}(\mu_1 H_s)^{3/2} + \frac{1}{4}\right]. \tag{4.55}$$

The contribution of the trapped modes to the total signal strength is given by the sum of the residues of integrand (2.40) in the poles $t_n = t_n^0 + \delta t_n$:

$$W(x, z, z_0) = \sqrt{\frac{\xi}{\pi}}\, e^{j\pi/4} \sum_{n=1}^{N} e^{j\xi t_n} F_n(h, h_0) \tag{4.56}$$

The function $F_n(h, h_0)$ is given by the following equations:
With $h, h_0 \leq H_s$,

$$F_n(h, h_0) = \frac{\mu_1}{\sqrt{\xi_n}} v(\mu_1 h - \xi_n) v(\mu_1 h_0 - \xi_n); \tag{4.57}$$

with $h > H_s$, $\quad h_0 \leq H_s$,

$$F_n(h, h_0) = \sqrt{\frac{\mu_1}{\mu_2}}\, \xi_n^{-1/4} \gamma_n^{1/2} w_1\left(\frac{t_n}{\mu_2^2} - \mu_2(h - H_s)\right) v(\mu_1 h_0 - \xi_n); \tag{4.58}$$

and with $h, h_0 > H_s$,

$$F_n(h, h_0) = \frac{\gamma_n}{\mu_1^2} w_1\left(\frac{t_n}{\mu_2^2} - \mu_2(h - H_s)\right) w_1\left(\frac{t_n}{\mu_2^2} - \mu_2(h_0 - H_s)\right). \tag{4.60}$$

For arbitrary heights and M-deficits of the evaporation duct the propagation constants t_n can be calculated by using numerical solutions of Eq. (4.47). In relation to a problem of quantum mechanics such a study was presented in Ref. [23] and associated references. In a problem of wave propagation through the troposphere a relevant study was presented in Ref. [24]. While the case of elevated M-inversion was considered in Ref. [24], it seems important that two branches of the propagation constant can be distinguished in that case. One branch corresponds to the "whispering gallery" modes, i.e. trapped modes, while the second branch corresponds to the diffraction modes. Figure 4.11 shows a similar study of the propagation constants t_n for an evaporation duct with a linear M-profile (4.46), where the circles show the position of the propagation constant in a complex plane t as a function of the duct height Z_s with constant gradients μ_1, μ_2. In this case we can also distinguish two branches of the propagation constants. As observed from Figure 4.2, the diffraction modes are concentrated near the ray $\arg(t) = \pi/3$, while the second branch of the roots of Eq. (4.47) traverses from the left half-plane of t to the right half-plane with increasing depth of the M-inversion, asymptotically reaching the positions corresponding to the trapped modes $t_n = \mu_1^3 H_s - \mu_1^2 \xi_n$, where $\xi_n = \left[3\pi/2\left(n - \frac{1}{4}\right)\right]^{2/3}$.
The waveguide mode with $n = 1$ is common to both branches. The contribution of the diffraction modes becomes negligible with increasing depth of the M-inversion

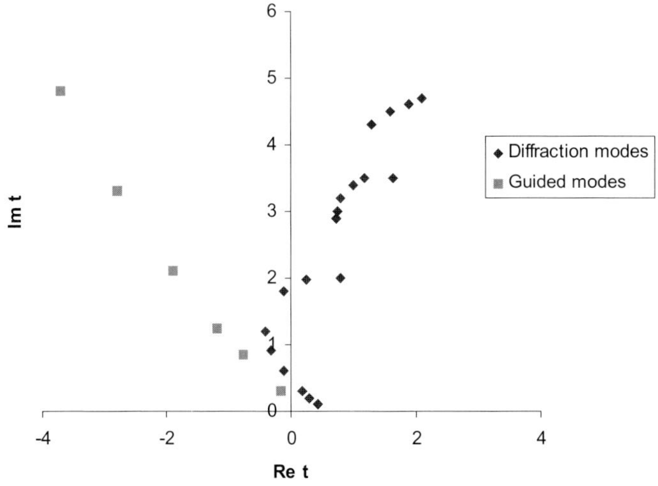

Figure 4.11 Evolution of the propagation constant in the t-plane with changes in the duct height.

compared with the contribution of the residues in the poles in the left hand-side of the t-plane.

We may also obtain the asymptotic solution for the characteristic equation (4.45) in the case of the smooth M-profile using Fock's theory [4]. In particular we are interested in the asymptote of Eq. (4.45) for small values of $\operatorname{Re} t_n$, i.e., in the vicinity of the minimum of the M-profile where it is most significantly different from the other approximations, especially the bilinear one. Using the WKB approximation for the height gain functions in Eq. (4.45) for a smooth but arbitrary profile we obtain

$$j\frac{\pi}{2}-\frac{\pi\nu}{2}-j\nu+j\nu\ln\nu+2jS_1-C_0+G=j2\pi n \tag{4.61}$$

where

$$C_0=0.91894;\quad G=\ln\Gamma\left(\frac{1}{2}+j\nu\right);\quad \nu=\frac{t_n}{\sqrt{2U''(H_s)}};$$
$$S_1=\frac{2}{3\mu_1^3}(U_0-t_n)^{3/2}-j\frac{\pi\nu}{2}. \tag{4.62}$$

For a linear-logarithmic M-profile under conditions of neutral stratification the parameters in Eq. (4.62) are defined as follows:
$U''(H_s)=\mu_2^3/H_s$ and $U_0=\mu_2^3 H_s\ln(H_s/H_r)$, where $H_r=kz_r/m$ and z_r is the minimum height at which the radio meteorological measurements can be carried out, the height z_r is normally associated with sea roughness.

The results of the calculation of the propagation constant t_1 of the first mode for both bilinear and linear-logarithmic M-profiles are shown in Table 4.1. The linear-logarithmic M-profile corresponds to the case of stable stratification when the value of the M-deficit ΔM and the duct height Z_s are related via the approximate expression: $\Delta M=g_s Z_s$, where $g_s=0.3$ N-units m^{-1}.

Table 4.1 Propagation constants of the first mode for bilinear and logarithmic M-profiles.

Duct height, m	Bilinear Model		Linear-logarithmic Model		Asymptotic
	Re t	Im t	Re t	Im t	Re t
20	6.992	0	6.994	0	6.998
19	6.237	0	6.27	0	6.268
18	5.529	0	5,562	0	5.559
17	4.838	0	4.871	0	4.867
16	4.167	0	4.199	0	4.195
15	3.515	0	3.548	0	3.453
14	2.887	0	2.919	9	2.914
13	2.285	0.0000467	2.316	0.00000154	2.309
12	1.717	0.00086	1.741	0.0000098	1.731
11	1.194	0.00857	1.199	0.000095	1.184
10	0.733	0.0452	0.7	0.0016	0.67
9	0.335	0.144	0.027	0.032	0.19
8	0.00002	0.33	-0.04	0.133	-0.023
7	–0.25	0.631	–0.043	0.46	–0.59
6	–0.85	1.04	–0.043	0.5	–0.89

The value of the M-deficit can be determined by approximating the logarithmic M-profile in the interval $0 < z < Z_s$ by a linear M-profile with gradient equal to $M'(z)\big|_{z=z_r}$, i.e., at the sea roughness level. As can be seen from comparison of the results in Table 4.1, for very strong M-inversion, when the mode is deeply trapped inside the duct, i.e., Re $>> t_n >> 1$, the propagation constants for the linear-logarithmic and bilinear profiles and asymptotical values are practically equal to each other and to asymptotical values for the linear profile $t_1^0 = \mu_1^3 H_s - \mu_1^2 2.338$, where $\mu_1^3 = a \cdot 10^{-6} g_s$. In the case of weak trapping, when Re $t_n \leq 1$, the difference between the two profiles becomes significant. This is caused by the behavior of the reflection of trapped waves from a potential barrier in the vicinity of the minimum of $U(h)$. In particular, the imaginary part of the propagation constant of the linear-logarithmic profile is less than one obtained from a linear profile since the thickness of the potential barrier is greater in the case of the logarithmic profile, Figure 4.12.

The contribution of the first mode in Eq. (4.46) in the case of the linear-logarithmic profile can be written in a form similar to Eqs. (4.57) to (4.60):

With $H_s << \mu_2^2 |t_1|^{1/4}$, then for all $h,\ h_0$

$$F_1(h, h_0) = \frac{1}{4\mu_2^2\sqrt{\xi_1}} \exp\left[jS(h, t_1) + jS(h_0, t_1) + j\frac{\pi}{3}\right]. \tag{4.63}$$

When $h,\ h_0 \leq H_s$

$$F_1(h, h_0) = -\frac{2D}{\Lambda} \sin[S(h, t_1)] \sin[S(h_0, t_1)] \tag{4.64}$$

where

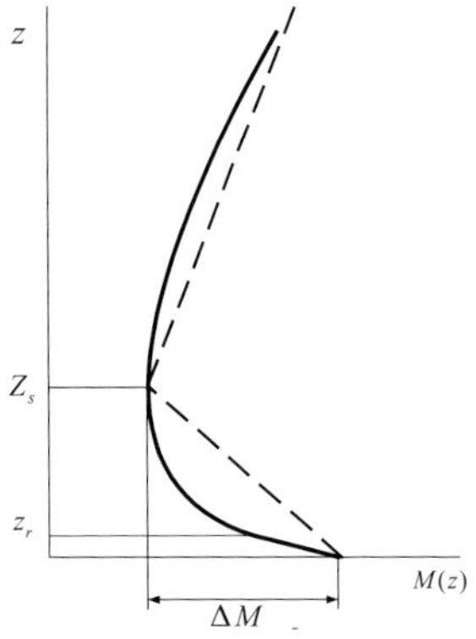

Figure 4.12 Smooth M-profile.

$$D = [(U(h) - t_1)\cdot(U(h_0) - t_1)]^{-1/4}, \Lambda = 2\frac{\sqrt{U_0 - t_1}}{\mu_1^3}.$$

When both antennas are located above the duct, i.e., $h,\ h_0 > H_s$, then

$$F_1(h, h_0) = \frac{T\cdot D}{2\Lambda}\exp[jS(h, t_1) + jS(h_0, t_1)] \tag{4.65}$$

where

$$T = 1 - |R_s(t_1)|^2,\ R_s = \frac{1}{1+e^{2\pi\nu t_1}}. \tag{4.66}$$

And, finally, when only one antenna is placed above the duct, say $h > H_s$, we obtain

$$F_1(h, h_0) = j\frac{D}{2\Lambda}\sin[S(h_0, t_1)]\exp[jS(h, t_1)]. \tag{4.67}$$

In the case of an arbitrary M-profile and a finite value of the surface impedance, the solution to Eq. (4.45) can be found by using numerical methods. In this case we may introduce a normalised height gain function f instead of the function $\chi^+(h, t)$

$$f(h, t) = \frac{\chi^+(h,t)}{\chi^+(0,t)}. \tag{4.68}$$

Function f is then governed by the uniform equation

$$\frac{d^2 f}{dh^2} + [U(h) - t]f = 0 \tag{4.69}$$

and satisfies the following boundary conditions at $h = 0$

$$f(0, t) = 1,\quad \left.\frac{df}{dh}\right|_{h=0} = q \tag{4.70}$$

where $q = jm/\sqrt{\eta}$.

Note that for horizontal polarization the parameter q should be replaced by $q' = jm\sqrt{\eta}$. At the upper boundary of the duct, at some height h_o, $h_o \geq H_s$, the function f should satisfy the condition of the outgoing wave, i.e., $d/dh\ \arg(f(h,t)) > 0$. While this height can be arbitrarily chosen, in practice the profile $U(h)$ is approximated by a linear one, thus allowing one to represent the outgoing wave via the Airy function w_1.

Equation (4.69) for the height-gain function f can be integrated by standard numerical methods making use of boundary conditions (4.70) and the analytical transition of the function and its derivative at the height h_o. The application of boundary conditions at both boundaries $h = 0$ and $h = h_o$ leads to the characteristic equation $Q(t_n) = 0$ for the propagation constants t_n.

The attenuation factor W in the shadow region can then be represented as the sum of normal waves equivalent to the residue sum of the integral (3.43) in the poles of the S-matrix:

$$W(\xi, h, h') = 2e^{j\pi/4}\sqrt{\pi\xi}\sum_{n=1}^{\infty} e^{j\xi t_n}\frac{dt_n}{dq} f(h, t_n) f(h', t_n) \tag{4.71}$$

where parameter $\frac{dt_n}{dq} = -\left[\frac{d}{dh}\left(\frac{\partial f}{\partial t_n}\right)\bigg|_{h=0}\right]^{-1}$ is called the excitation coefficient of the normal wave with number n.

The characteristic equation $Q(t_n) = 0$ in the general case can be realised in computer-based calculations by simultaneously solving Eq. (4.69) for height-gain functions with a trial value of t_n. In the case of the bilinear profile (4.46) the characteristic equation $Q(t_n) = 0$ is exactly the same as Eq. (4.47). In the case of the linear-logarithmic profile, as shown in Section 4.1, the refractivity profile close to the sea surface can be replaced at the heights $h < h_r$ by a linear segment with appropriate gradient without appreciable errors in the estimation of the propagation constant. In that case the resulting profile consists of three segments: the linear profile in the interval $0 \leq h < h_r$, the logarithmic profile $U(h) = U(h_r) - \kappa \ln(\eta/h_r)$, and the linear profile at heights $h > h_o$. In this case the solution to Eq. (4.52) should be tailored in terms of the continuity of the function f and its derivative df/dh with analytical representation of the function f via Airy functions in linear segments.

The numerical algorithm for finding the propagation constants can be implemented in the following way. Assume that we know the etalon solutions $\overset{0}{t_n}$ for a specific profile $U_0(h)$. The respective characteristic equation we denote as $Q(\overset{0}{t_n})$. Instead of the equation $Q(t_n) = 0$ for the propagation constant of the problem we may use

$$Q(t_n, \sigma) = (1 - \sigma)Q(\overset{0}{t_n}) \tag{4.72}$$

introducing a new parameter σ varying between 0 and 1. The roots of Eq. (4.72) depend on thevalue of σ. When $\sigma = 0$ we have a solution of the etalon problem, i.e. $t_n = \overset{0}{t_n}$, with $\sigma = 1$ the solution to Eq. (4.72) gives "true" propagation constants. With variation in σ from 0 to 1 the roots of Eq. (4.72) traverse along some trajectories in the complex plane t from $\overset{0}{t_n}$ to t_n. The equation for the trajectories is

$$\frac{dt_n}{d\sigma} = -\frac{Q(t_n^0)}{\dfrac{\partial}{\partial t_n} Q(t_n, \sigma)}. \tag{4.73}$$

In practice, the parameter σ should be related to one of the inherent parameters of the problem, i.e., this might be the impedance q, the height of the duct H_s or the gradient of the M-profile above or below H_s. In that case the trajectories of the roots $t_n(\sigma)$ are likely to be restrained by their own valley in the module profile of $|Q(t_n, \sigma)|$ in the plane of complex t [7]. The algorithm based on the evolution of the height of the evaporation duct has been implemented in a radio coverage prediction system [25, 26] with the close participation of the author. In that system we used a bilinear approximation of the linear-logarithmic M-profile for the evaporation duct described in terms of a single parameter, the height of the duct, H_s. The approximating linear profile reveals an almost constant gradient within $h < H_s$ thus allowing one to use one parameter H_s to build the evolution algorithm similar to Eq.(4.55) and create look-up tables for pre-calculated propagation constants $t_n(H_s)$.

Various well-developed methods of calculation of the propagation constants can be found in the literature. The formalism of the duct propagation in a stratified troposphere has been developed in a number of classical works started in the 1940s [6, 27, 28]. Until now, those publications provide a benchmark reference and a framework for the detailed study of duct propagation. The practical algorithms incorporating the mode theory have been implemented in various computer-based programs and prediction systems [29, 30]. Among known models of the M-profile describing the average refractivity in the atmospheric boundary layer, the linear-logarithmic profile is accepted as the most adequate M-profile model for the evaporation duct to be utilised in development of the radar coverage prediction systems.

4.4 Coherence Function in a Random and Non-uniform Atmosphere

4.4.1 Approximate Extraction of the Eigenwave of the Discrete Spectrum in the Presence of an Evaporation Duct

Consider Eq. (2.35) in the case when the evaporation duct is present in the height interval $(0, H_s)$. Let us seek a solution to the attenuation function W in the expansion over the system of eigenfunctions of the continuous spectrum defined in Section 2.2

$$W(\vec{r}) = \int_{-\infty}^{\infty} dE \cdot A(x, y, E)\Psi_E(z), \tag{4.74}$$

$$A(x, y, E) = \int_{0}^{\infty} dz \cdot W(x, y, z)\Psi_E^*(z). \tag{4.75}$$

The function $\Psi_E(z)$ obeys Eq. (3.37), boundary conditions (2.38) and conditions (2.40), (2.41) for orthogonality and completeness respectively. It also has singularities of pole type in an upper half-space of E, asymptotically approaching ray $\arg E = \pi/3$ with $|E| \to \infty$ when $\varepsilon_m(z) \to 2\,z/a$ with $z \to \infty$. Because of the presence of M-inversion in a near-surface layer, an evaporation duct, some finite number of poles E_n, $(n = 1, 2, \ldots N)$ lie close to the real axis of E. The value of Re E_n, denoted further as $\tilde{E}_n$, belongs to the interval $k^2(\varepsilon_m(H_s) - 1) < \tilde{E}_n < k^2(\varepsilon_m(0) - 1)$. Such poles correspond to waves trapped in the waveguide channel created by the evaporation duct. Without loss of generality we consider here the situation when $N = 1$, i.e. a single mode waveguide (evaporation duct). It is the most common case for the evaporation duct propagation of radio waves in a range of about 10 GHz.

Let us split the interval of the integration in Eq. (4.49) into two domains and set $W = W_1 + W_2$ where

$$\begin{aligned} W_1(\vec{r}) &= \int_{E_m}^{\infty} dE \cdot A(x, y, E)\Psi_E(z), \\ W_2(\vec{r}) &= \int_{-\infty}^{E_m} dE \cdot A(x, y, E)\Psi_E(z), \end{aligned} \tag{4.76}$$

and $E_m = k^2(\varepsilon_m(H_s) - 1)$. Using Eq. (4.75) we can evaluate the contribution of the term W_1. In the shadow region $x > \sqrt{2a}(\sqrt{z} + \sqrt{z_0})$, where z, z_0 are the heights of the transmitting and receiving antennas respectively, the major contribution to the integral (4.51) comes from the E-interval $k^2(\varepsilon_m(H_s) - 1) < E < k^2(\varepsilon_m(0) - 1)$ and the height interval $z < H_s$:

$$\begin{aligned} W_1 &= \int_{E_m}^{\infty} dE \cdot \int_0^{\infty} dz'\, W(x,y,z')\Psi_E^*(z')\Psi_E(z) \approx \\ &\int_{E_m}^{E_0} dE \cdot \int_0^{H_s} dz'\, W(x,y,z')\Psi_E^*(z')\Psi_E(z). \end{aligned} \tag{4.77}$$

Within the interval $E_m < E < E_0$ we can distinguish a function $\phi_E^0(z)$, which depends little on E

$$\Psi_E(z) \approx C(E)\phi_E^0(z) \tag{4.78}$$

and normalised by the following equation

$$\int_0^{Z_s} \phi_E^0(z)\phi_E^{0*}(z)dz = 1. \tag{4.79}$$

The meaning of Eqs. (4.78) and (4.79) is that we approximated the function $\Psi_E(z)$ of the continuous spectrum by the eigenfunction of discrete spectrum $\phi_E^0(z)$ localised inside the evaporation duct in such a way that the residue of Eq. (4.74) in the pole $E = \tilde{E}_n$ would approximately provide the contribution of the trapped wave.

Using Eq. (4.78) the term W_1 is given by

$$W_1 = \int_{E_m}^{E_0} dE \cdot |C(E)|^2 \phi_E^0(z) \int_0^{H_s} dz' W(x, y, z')\phi_E^{0*}(z'). \tag{4.80}$$

Now assuming that $\varepsilon_m(z) = 2\, z/a$ for $z > H_s$ we obtain

$$\Psi_E(z) = \frac{1}{2\sqrt{\pi\mu}}\left[w_2\left(\frac{E}{\mu^2} - \mu(z - H_s)\right) - S(E)w_1\left(\frac{E}{\mu^2} - \mu(z - H_s)\right)\right] \tag{4.81}$$

where $\mu = k/m$. The coefficient $S(E)$ is determined by combining the solution (4.81) for a large height with the boundary conditions at the surface $z = 0$. In the vicinity of the pole $E_n = \tilde{E}_n + j\delta_n$ the coefficient $S(E)$ has the form [22]:

$$S(E) \approx B(E)\frac{E - \tilde{E}_n + j\delta_n}{E - \tilde{E}_n - j\delta_n} \tag{4.82}$$

where $B(E)$ is a slowly varying function of E. Returning to Eq. (4.80) and using Eqs. (4.82), (4.79) and (4.81) we obtain an explicit expression for $|C(E)|^2$:

$$|C(E)|^2 = -2j\frac{dS^*}{dE}S + S^*\left[\left[\Psi'_E(Z_s)\right]^2 - E\Psi_E^2(Z_s)\right]. \tag{4.83}$$

Retaining the resonant term $2j\dfrac{dS^*}{dE}S$ in Eq. (4.58) as $\delta_n \to 0$, we obtain

$$|C(E)|^2 = \frac{\delta_n}{\left(E - \tilde{E}_n\right)^2 + \delta_n^2}. \tag{4.84}$$

Since the function (4.84) has a sharp maximum at $E = \tilde{E}_n$ and

$$\int_{E_m}^{\infty} |C(E)|^2 dE \approx \int_{-\infty}^{\infty} |C(E)|^2 dE = \pi, \tag{4.85}$$

function (4.84) can be approximated by a delta function

$$|C(E)|^2 \approx \delta(E - E_n). \tag{4.86}$$

Then, substituting Eq. (4.86) into Eq. (4.80) we obtain for $W(\vec{r})$

$$W(x, y, z) \cong \phi_{E_1}^0(z) \int_0^{\infty} W(x, y, z')\phi_{E_1}^{0*}(z')dz' + \int_{-\infty}^{E_m} A(x, y, E)\Psi_E(z). \tag{4.87}$$

As observed from Eq. (4.87), the contribution of the region of $E > E_m$ into integral (4.74) is represented by expansion over the eigenfunctions of the discrete spectrum $\phi_{E_n}^0$ that actually represent the boundary problem (2.37), (2.38) with term $2\, z/a$ in $\varepsilon_m(z)$ equal to zero. The function $\phi_{E_n}^0(z)$ coincides, up to exponentially small terms, with the real eigenfunction $\phi_1(z)$ of the discrete spectrum in the height region $z \ll \mu^2/\delta_1^2$. Therefore, for consistency of the approximation made, we have to exclude scattering in the troposphere's layer at the height $z \gg \mu^2/\delta_1^2$. For a single scattering it results in the limitation the distance

$$x << 2k/\delta_1. \tag{4.88}$$

For frequencies of about 10 GHz and evaporation duct heights H_s of the order of 10–15 m, the value of δ_1 usually does not exceed $10^{-3}(k/m)^2$, in which case the inequality (4.63) is satisfied at the distances $x << 10^4$ km.

4.4.2
Equations for the Coherence Function

Lets consider the equations for the coherence function

$$\Gamma(x, \vec{\rho}_1, \vec{\rho}_2) = \frac{\Gamma_w(x, \vec{\rho}_1, \vec{\rho}_2)}{a^2} \tag{4.89}$$

and $\Gamma_w(x, \vec{\rho}_1, \vec{\rho}_2) = \langle W(x, \vec{\rho}_1) \cdot W^*(x, \vec{\rho}_2)\rangle$, where $\vec{\rho}_1 = \{y_1, z_1\}$, $\vec{\rho}_2 = \{y_2, z_2\}$ are the coordinates of the observation points at the distance x from the source. The closed equation, similar to Eq. (2.68), for the coherence function can be obtained for $\Gamma_w(x, \vec{\rho}_1, \vec{\rho}_2)$:

$$\frac{\partial \Gamma_w}{\partial x} - \frac{j}{2k}(\Delta_{\perp 1} - \Delta_{\perp 2})\Gamma_w - \frac{\pi k^2}{4} H(\vec{\rho}_1 - \vec{\rho}_2)\Gamma_w = 0 \tag{4.90}$$

where

$$\Delta_{\perp i} = \frac{\partial^2}{\partial y_i^2} + \frac{\partial^2}{\partial z_i^2} + k^2(\varepsilon_m(z_i) - 1), \quad i = 1, 2, \tag{4.91}$$

and $H(\vec{\rho})$ is a structure function of the fluctuations in $\delta\varepsilon$. The conditions of applicability of Eq. (4.90) are similar to those listed in Section 2.3 and are given by a set of inequalities:

$$\text{a)}\ (kH_s)^2 >> 1, \qquad \text{b)}\ (kL_\perp)^2 >> 1,$$
$$\text{c)}\ \frac{kH_s^2}{L_x} >> 1, \qquad \text{d)}\ \frac{kL_\perp^2}{L_x} >> 1. \tag{4.92}$$

The inequality (4.92 c) means de-correlation of the consecutive acts of scattering of the wave: the interval of the longitudinal correlation L_x should be less than the length of the cycle Λ in the waveguide, $\Lambda \sim H_s/\vartheta_c$, and $\vartheta_c \sim 1/(kH_s)$ is a characteristic sliding angle of the trapped wave. The condition (4.92 (d)) means that the Fresnel-zone size $\sqrt{\lambda L_x}$ is less than the vertical scale of the inhomogeneities $L_\perp$, i.e. between consecutive acts of scattering, the distance between which is of the order of L_x, the wave propagates as in a uniform medium, and we do not take into account diffraction at the inhomogeneities of $\delta\varepsilon$.

As follows from Eq. (4.87), the coherence function $\Gamma_w(x, \vec{\rho}_1, \vec{\rho}_2)$ can be presented as the superposition of the discrete and continuum eigenfunctions:

$$\Gamma_w(x,\vec{\rho}_1,\vec{\rho}_2) = g_d(x,y_1,y_2)\phi_d(z_1)\phi_d(z_2)+ \\ 2\mathrm{Re}\int_{-\infty}^{E_m} dE\cdot g_{cd}(E,x,y_1,y_2)\Psi_E(z_1)\phi_d(z_2)+ \\ \int_{-\infty}^{E_m} dE_1\cdot\int_{-\infty}^{E_m} dE_2\cdot g_c(E_1,E_2,x,y_1,y_2)\Psi_{E_1}(z_1)\Psi^*_{E_2}(z_2). \tag{4.93}$$

In Eq. (4.93) the term g_d determines the part of the coherence function carried by trapped modes, g_c is contributed by the waves of the continuum spectrum, and g_{cd} is a term responsible for the combined mechanism of transfer of the coherence function.

Substituting Eq. (4.93) into Eq. (4.90), introducing the variables: $Y = (y_1 + y_2)/2$ and $y = y_1 - y_2$ and the Fourier transform of g_d over variable Y:

$$g_d(x,y_1,y_2) = \int_{-\infty}^{\infty} dp\cdot\tilde{g}_d(x,p,y)e^{-jpY}. \tag{4.94}$$

For the other terms, g_c and g_{cd}, the Fourier transforms are introduced in similar way. Let us also introduce the notations

$$\gamma_0 = \frac{\pi k^2}{2}\int d^2\vec{\kappa}_\perp\Phi_\varepsilon(0,\vec{\kappa}_\perp),\quad \eta = y + p\frac{x}{k},\quad D(\kappa_z,\eta) = \Phi_\varepsilon(0,\vec{\kappa}_\perp)e^{j\kappa\eta_y},$$

we obtain the system of coupled equations, which describes the energy exchange between the trapped waves and the waves of the continuous spectrum due to a scattering on the fluctuations of $\delta\varepsilon$:

$$\frac{\partial\tilde{g}_d(x,p,\eta)}{\partial x} + \gamma_0\tilde{g}_d(x,p,\eta) = \int d^2\vec{\kappa}_\perp D(\vec{\kappa}_\perp,\eta)\times \\ \left\{\begin{array}{l} \tilde{g}_d(x,p,\eta)|V_{11}(\kappa_z)|^2 + \int_{-\infty}^{E_m} dE_1\tilde{g}_{cd}(x,E_1,p,\eta)V_{11}(\kappa_z)\cdot(V_1^+(E_1,\kappa_z))^* + \\ \int_{-\infty}^{E_m} dE_2\tilde{g}_{cd}(x,E_2,p,\eta)V_{11}^*(\kappa_z)\cdot V_1^-(E_1,\kappa_z)+ \\ \int_{-\infty}^{E_m} dE_1\int_{-\infty}^{E_m} dE_2\tilde{g}_c(x,E_1,E_2,p,\eta)V_1^+(E_1,\kappa_z)V_1^-(E_2,\kappa_z) \end{array}\right\}, \tag{4.95}$$

$$\frac{\partial\tilde{g}_{cd}(x,E,p,\eta)}{\partial x} + \left[\gamma_0 - \frac{j}{2k}(E_1 - E)\right]\tilde{g}_{cd}(x,E,p,\eta) = \int d^2\vec{\kappa}_\perp D(\vec{\kappa}_\perp,\eta)\times \\ \left\{\begin{array}{l} \tilde{g}_d(x,p,\eta)V_{11}(\kappa_z)V_1^-(E,\kappa_z)+ \\ \int_{-\infty}^{E_m} dE_1\tilde{g}_{cd}(x,E_1,p,\eta)\cdot V_{11}(\kappa_z)V^*(E_1,E,\kappa_z)+ \\ \int_{-\infty}^{E_m} dE_2\tilde{g}_{cd}(x,E_2,p,\eta)\cdot V_1^-(E,\kappa_z)V_1^+(E_1,\kappa_z)+ \\ \int_{-\infty}^{E_m} dE_1\int_{-\infty}^{E_m} dE_2\tilde{g}_c(x,E_1,E_2,p,\eta)V_1^+(E_1,\kappa_z)V^-(E_2,E,\kappa_z) \end{array}\right\}, \tag{4.96}$$

$$\frac{\partial \tilde{g}_c(x,E,E',p,\eta)}{\partial x} + \left[\gamma_0 - \frac{j}{2k}\left(E - E'\right)\right]\tilde{g}_c(x,E,E',p,\eta) = \int d^2\vec{\kappa}_\perp D(\vec{\kappa}_\perp,\eta) \times$$

$$\left\{\begin{array}{l} \tilde{g}_d(x,p,\eta)(V_1^-(E,\kappa_z))^* V_1^-(E',\kappa_z) + \\ \int\limits_{-\infty}^{E_m} dE_1 \tilde{g}_{cd}(x,E_1,p,\eta)\cdot(V_1^-(E',\kappa_z))^* V^*(E_1,E',\kappa_z) + \\ \int\limits_{-\infty}^{E_m} dE_2 \tilde{g}_{cd}(x,E_2,p,\eta)\cdot V(E_2,E,\kappa_z)V_1^-(E',\kappa_z) + \\ \int\limits_{-\infty}^{E_m} dE_1 \int\limits_{-\infty}^{E_m} dE_2 \tilde{g}_c(x,E_1,E_2,p,\eta)V(E_1,E,\kappa_z)V^*(E_2,E',\kappa_z) \end{array}\right\}. \quad (4.97)$$

The functions

$$V_{11}(\kappa_z) = \int\limits_0^\infty dz e^{j\kappa_z z}\phi_d(z)\phi_d^*(z), \quad (4.98)$$

$$V_1^\pm(E,\kappa_z) = \int\limits_0^\infty dz e^{\pm j\kappa_z z}\phi_d(z)\Psi_E(z), \quad (4.99)$$

$$V(E_1,E_2,\kappa_z) = \int\limits_0^\infty dz e^{j\kappa_z z}\Psi_{E_1}(z)\Psi_{E_2}^*(z) \quad (4.100)$$

are the coefficients of scattering on inhomeogeneties $\delta\varepsilon$ with vertical scales $l_z = 2\pi/\kappa_z$ for waveguide modes (4.98), waveguide modes into the waves of the continuous spectrum (4.74), and continuum into continuum (4.100). The relative value of the contribution of each of the terms $\tilde{g}_d$, $\tilde{g}_{cd}$, and $\tilde{g}_c$ to the total field at the observation point depends on the form of the initial distribution, $\Gamma(x=0,\vec{\rho}_1,\vec{\rho}_2)$, and the position of the receiver relative to the evaporation duct.

It is reasonable to assume, when the points of observations are located inside the evaporation duct, $z_1, z_2 < H_s$, that the major contribution to the coherence function comes from the trapped waves. Therefore, in a first order approximation, we can neglect the contribution from g_c and g_{cd} to g_d and the total coherence function Γ_w, considering the multiple scattering of the trapped waves only. In this case we obtain

$$\frac{\partial \tilde{g}_d}{\partial x} + \left[\gamma_s - \int d^2\vec{\kappa}\cdot D(\vec{\kappa},\eta)|V_{11}(\kappa_z)|^2\right]\tilde{g}_d = 0. \quad (4.101)$$

Substituting the solution to Eq. (4.101) into Eq. (4.93), we can use the orthogonality feature between the waves of the discrete and continuous spectra. Performing the Fourier transform, inverse to Eq. (4.94), we obtain

$$\Gamma_w(x,y,Y,z_1,z_2) = \frac{k}{2\pi x}\phi_d(z_1)\phi_d(z_2)\int\limits_{-\infty}^{\infty} dy' \int\limits_{-\infty}^{\infty} dY' \Gamma_w(x=0,y',Y') \times \exp\left[j\frac{k}{x}(y-y')(Y-Y') - P(x,y',y)\right]. \quad (4.102)$$

The parameter $P(x, y', y)$,

$$P(x,y',y) = \frac{\pi k^2}{2} \int d^2\vec{\kappa}_\perp \Phi_\varepsilon(0,\vec{\kappa}_\perp) \times$$
$$\left\{ x - |V_{11}(\kappa_z)|^2 \int_0^x dx' \exp\left[j\kappa_y \left(y\frac{x'}{x} + y'\left(1 - \frac{x'}{x}\right)\right)\right]\right\}, \quad (4.103)$$

after integration over dy' determines the attenuation exponent of the coherence function with distance x. The first term in Eq. (4.103) describes the attenuation of the average field due to energy transfer to the incoherent component. The second term determines the incoherent contribution of the energy scattered back to the waveguide mode in the direction of propagation.

Let us consider the intensity of the field at the point $\vec{r} = \{x, 0, z\}$ produced by a point source located at $\vec{r}_0 = \{0, 0, z_0\}$. The initial distribution of the field in Eq. (4.77) takes the form:

$$\Gamma_w\left(x = 0, y', Y'\right) = \frac{a^2}{4k^2} \phi_d^2(z_0) \delta\left(y'\right) \delta\left(Y'\right). \quad (4.104)$$

Now, we can assume that $\varepsilon_m(z)$ is described by a bilinear function with gradient $\nu = d\varepsilon_m/dz$, for $z < H_s$. While computation of Eq. (4.96) can be performed with any regular function $\varepsilon_m(z)$, the bilinear approximation provides an analytical solution useful for qualitative analysis of the scattering mechanism. Introducing the parameters:

$\mu_1 = a|\nu|/2$, $H_s = kZ_s/m$, $h = k\xi/m$, $h_0 = kz_0/m$, and $\tau_1 = -\mu_1^2 2.338 + \mu_1^3 H_s$ we obtain a solution for the intensity of the field, normalised on the intensity in a free space:

$$J = \frac{x^2}{a^2} |W(x,\vec{r})|^2 = \frac{x^2}{a^2} \Gamma_w(x,0,z) = \frac{kx}{8\pi m^2 \tau_1} \upsilon^2 \left(\frac{\tau}{\mu_1^2} - \mu_1 (H_s - h_0) \right) \times$$
$$\upsilon^2 \left(\frac{\tau}{\mu_1^2} - \mu_1 (H_s - h) \right) \exp\left(-\gamma_d x\right). \quad (4.105)$$

Here, the function ϕ_d is expressed via the Airy function $\upsilon(x)$ of the real argument since the pole E_1 is regarded as a real one, $\mathrm{Im}\{E_1\} = 0$. The attenuation exponent in Eq. (4.105) is given by

$$\gamma_d = \frac{\pi k^2}{2} \int d^2\vec{\kappa}_\perp \Phi_\varepsilon(0,\vec{\kappa}_\perp)[1 - |V_{11}(\kappa_z)|] \quad (4.106)$$

where $\vec{\kappa}_\perp = \{\kappa_y, \kappa_z\}$.

Consider the calculation of γ_d with the spectrum given by Eq. (1.36) for locally uniform anisotropic fluctuations $\delta\varepsilon$ and introduce the non-dimensional variable $t = z / Z_0$, where $Z_0 = m_0 / k$, the characteristic scale of the variations in the function $\phi_d(z)$, $m_0 = (k/|\nu|)^{1/3}$. Performing the integration over κ_y and introducing the variable $q = \kappa_z Z_0$, we obtain

$$\gamma_d = 0.17k^2 C_{\varepsilon\perp}^2 \left(\frac{Z_0}{\alpha}\right)^{5/3} A \tag{4.107}$$

where A is a constant of the order of unity, the exact value of which is defined by a true behavior of the height function $\varphi_d(z)$:

$$A = \int_0^\infty dq \cdot q^{-8/3} \left[1 - \left|\int_0^\infty dt \cdot e^{jqt} \varphi_d^2(t)\right|\right]^2 . \tag{4.108}$$

In the case of the bilinear model of $\varepsilon_m(z)$, Eq. (4.108) takes the form

$$A = \int_0^\infty dq \cdot q^{-8/3} \left[1 - \frac{1}{\tau_1^2}\left|\int_0^\infty dt \cdot e^{jqt} v^2(t-\tau_1)\right|\right]^2 , \tag{4.109}$$

and calculation of Eq. (4.107) results in the value $A = 1.51$. Hence, for γ_d we obtain finally

$$\gamma_d = 0.264k^{8/9} C_{\varepsilon\perp}^2 \alpha^{-5/3} |\nu|^{-5/9} . \tag{4.110}$$

Equation (4.110) is valid for locally uniform turbulent fluctuation $\delta\varepsilon$, even when the external scales of the turbulence are infinite: $L_z, L_\perp \to \infty$, $\sigma_\varepsilon^2 \to \infty$. As discussed in Chapter 1, real measurements of the fluctuations $\delta\varepsilon$ are always limited in time, and for the purpose of comparison with experiment another model of spectrum (1.35), with finite values of σ_ε^2 and external scales, can be used instead of Eq. (1.36). The match between models is achieved when $\sigma_\varepsilon^2 = 1.9C_{\varepsilon\perp}^2 L_\perp^{2/3}$.

The calculation of Eq. (4.106) can be simplified when $L_z << Z_s$. In this case, the second term can be neglected and the attenuation factor is entirely determined by the attenuation of the coherent component of the wave field:

$$\gamma_d \approx \gamma_S = \frac{\pi \kappa^2}{2} \int d^2\kappa_\perp \Phi_\varepsilon(0, \vec{\kappa}) = 0.374\sigma_\varepsilon^2 k^2 L_\perp . \tag{4.111}$$

The apparent reason for this is that the scattering on the small-scale fluctuations, with the scattering angle greater than the critical angle of the waveguide $\vartheta_c \sim 1/(kH_s)$, leads only to a flow of energy from the waveguide.

Until now we have assumed that the fluctuations in $\delta\varepsilon$ were statistically uniform over the height over the surface, i.e., parameters C_ε, α as well as L_z do not depend on the height z. However, from the theory of atmospheric turbulence [31], it follows that the external vertical scale of the fluctuations can be regarded as a linear function of height: $L_z = \beta z$, where β is a coefficient with numerical value less than unity. To some extent, a quasi-uniform behavior of the fluctuations $\delta\varepsilon$ can be accounted for by using the values of C_ε and α at the height z_m, where the scattering is more intense, i.e. at the point of the maximum of the first normal wave φ_d. In the case of the bilinear approximation, $z_m = 1.32\ m_0/k$. Thus,

$$\alpha = \frac{\beta z_m}{L_x} = \frac{1.32\beta}{k^{2/3}} |\nu|^{1/3} L_x \tag{4.112}$$

and, instead of Eq. (4.110), we obtain

$$\gamma_d = 0.166 k^2 C_{\varepsilon\perp}^2 L_x^{5/3} \beta^{-5/3}. \tag{4.113}$$

In fact, the attenuation factors, defined by both Eqs. (4.113) and (4.111), will be equal since the vertical scale of the fluctuations $\delta\varepsilon$ will be less than the thickness of the evaporation duct, $L_z << Z_s$. From comparison with Eqs. (4.113) and (4.111), the value of β can be estimated as $\beta = 0.4$ which agrees well with the measurement, and, therefore, Eq. (4.86) can be used to estimate the attenuation of the radio wave in an evaporation duct. Figure 4.13 [32] shows some results of a comparison of the field strength J in the evaporation duct relative to one in a free space at the frequency 10 GHz.

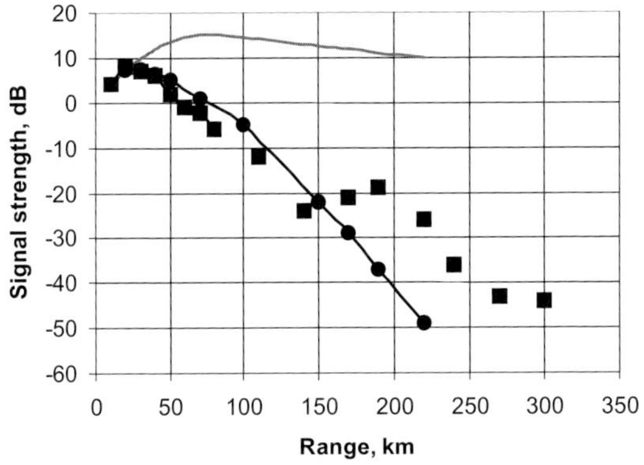

Figure 4.13 Signal strength in an evaporation duct at 10 GHz:
■ Measured signal; ● Calculated, $\delta\varepsilon \neq 0$; – Calculated, $\delta\varepsilon = 0$.

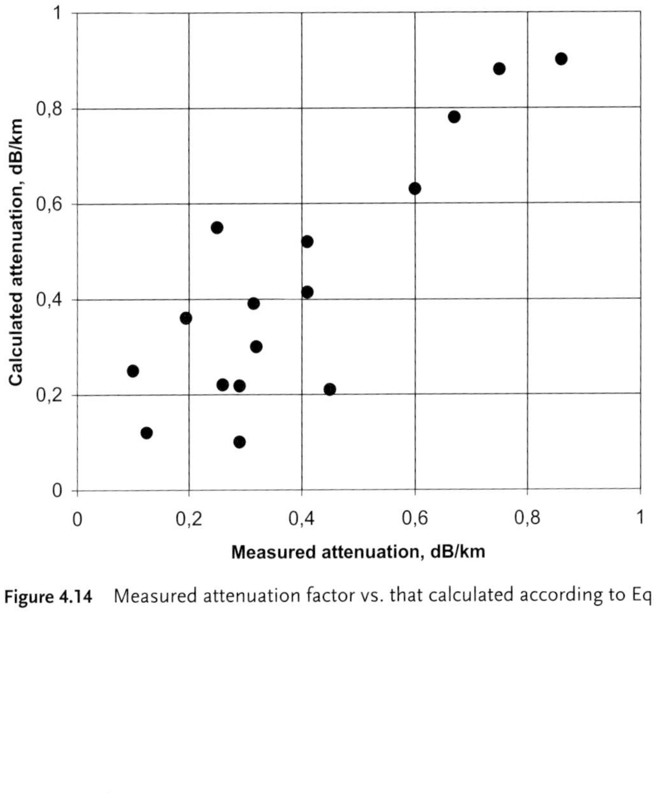

Figure 4.14 Measured attenuation factor vs. that calculated according to Eq. (4.111) [32].

As observed from Figure 4.13, the calculation of the field strength in an evaporation duct using only the mean M-profile, when $\delta\varepsilon = 0$, provides unrealistically high levels of the signal beyond the horizon. The curve marked by the squares is calculated with Eq. (4.80) using Eq. (4.86). The measurements of the fluctuation $\delta\varepsilon$ performed at the time of the radio measurements provided the following values: $C_N = 0.09$ *N*-units cm$^{-1/3}$, $L_\perp = 48$ m at a distance less than 100 km; and $C_N = 0.11$ *N*-units cm$^{-1/3}$, $L_\perp = 52$ m at a distance in the range between 100 and 200 km. Here $C_N = 1/2\ 10^6 C_\varepsilon$.

Figure 4.14 from Ref. [32] shows results of a comparison between the measured γ_m and the calculated (according to the theory provided in this section) attenuation factors γ_d at the frequency 10 GHz in the presence of an evaporation duct over the ocean. The data were collected from 16 tests when radio measurements were performed synchronously with refractometer measurements of the fluctuations in the near-surface layer of the troposphere. The correlation coefficient between the measured and calculated values of the attenuation factor is 0.8, which suggests that the wave scattering at the fluctuations of $\delta\varepsilon$ provides a significant contribution to the mechanism of the radio wave propagation in an evaporation duct.

4.5 Excitation of Waves in a Continuous Spectrum in a Statistically Inhomogeneous Evaporation Duct

In the absence of fluctuations in refractive index, $\delta\varepsilon$, the signal strength in the shadow region above the evaporation duct is exponentially small because of the smallness of the "sub-barrier leakage" of the trapped modes. The contribution of the waves of the continuous spectrum, initially excited by the transmitter, can be neglected because of the diffraction attenuation in the shadow region. The scattering of the waveguide trapped modes by random non-uniformities of the refractive index gives rise to incoherent exchange of energy between trapped modes and excitation of waves in the continuous spectrum. In addition, the initially excited waves of the continuous spectrum may reach beyond the horizon due to a scattering in the upper layers of the troposphere (Booker–Gordon's single scattering mechanism). In this section, we study the contribution of the waves of the continuous spectrum, whose source is a waveguide mode, to the intensity of the field, i.e. the mechanism of the excitation of waves of the continuous spectrum by trapped modes of the evaporation duct scattered on the imhomogeneities in the refractive index.

Recalling the equations for the coherence function from the previous section, we concentrate on investigation of the spectral amplitude of the component of the coherence function related to a continuous spectrum:

$$\tilde{g}_c(x, E_1, E_2, p, y) = \frac{1}{2\pi} \int_0^\infty dz_1 \int_0^\infty dz_2 \int_{-\infty}^{\infty} dY \cdot \Gamma(x, Y, y, z_1, z_2) \Psi^*_{E_1}(z_1) \Psi_{E_2}(z_1) \exp(jpY) \tag{4.114}$$

As is well known [31], in the presence of random fluctuations of the refractive index, $\delta\varepsilon$, the coherent component of the waves of the continuous spectrum attenuates with distance due to incoherent multiple scattering. The decrement of attenuation in the coherent component is given by

$$\gamma_c = \frac{\pi k^2}{2} \int d^2\vec{\kappa}_\perp \Phi_\varepsilon(0, \vec{\kappa}_\perp) \tag{4.115}$$

where $\Phi_\varepsilon(0, \vec{\kappa}_\perp)$ is the spectrum density of the fluctuations in $\delta\varepsilon$. As observed in Section 2.3, in the case of free-space propagation with random irregularities of $\delta\varepsilon$, the attenuation of the coherent component in the field intensity is completely compensated by the inflow of the incoherently scattered field from other directions, and the total intensity of the wave field does not reveal exponential attenuation. Since the propagation above the evaporation duct is free-space-like we may use the above argument in support of neglecting multiple scattering of the wave of the continuous spectrum in a space above the evaporation duct. However, the effect of multiple scattering of the wave of the continuous spectrum can obviously be neglected if we demand a smallness of the attenuation of the coherent component of the CS spectrum wave, i.e., $\gamma_c x_a << 1$, over the path x_a of the wave from the point where scattering of the wave guide mode into the CS wave occurred up to the point of observation. Evaluating the maximum value of the distance x_a as $x_{a\max} = \sqrt{2az}$, where z is the height of the observation point, we obtain the limitation $z << \left(2a\gamma_c^2\right)^{-1}$. In the case of Kolmogorov's turbulence $\gamma_c = 2.66k^2 C_n^2 L_0^{5/3}$. Assuming C_n^2=10^{-15} cm$^{-2/3}$, $L_0 = 10^4$ cm, $k = 2$ cm^{-1}, we obtain the result that the maximum heights under consideration should not exceed the value $z_{\max} \sim 10^3$ m. Thus, the approximation in which the estimate is obtained below for the waves of the continuous spectrum is applicable for altitudes of the receiving point z satisfying the inequality $z \le z_{\max}$.

Let us study the intensity of the field of the waves of the continuous spectrum $I_c(x, z) = \Gamma(x, \vec{\rho}, \vec{\rho})$, $\vec{\rho} = \{0, z\}$, setting the y-coordinates of the receiving and transmitting antennas to zero. Then, integrating Eq. (4.97) and employing the expression (4.104) for the component of the coherence function associated with waves of the discrete spectrum, Γ_d, we obtain

$$I_c(x, z) = \frac{ka^2}{16x} \left|\phi_d(z_0)\right|^2 \exp\left(-\gamma_d x\right)$$
$$\times \int_0^x dx' \int_{-\infty}^{\infty} dq \int_{Q<-|q|/2} dQ B(q, Q) \Psi_{Q+q/2}(z) \Psi^*_{Q-q/2}(z) \exp\left(j\frac{qx'}{2k} + \gamma_d x'\right) \tag{4.116}$$

where we have introduced $q = E_1 - E_2$, $Q = (1/2)(E_1 + E_2)$,

$$B(q, Q) = \int_{-\infty}^{\infty} d\kappa_y \int_{-\infty}^{\infty} d\kappa_z \Phi_\varepsilon(0, \kappa_y, \kappa_z) \times$$
$$\int_0^\infty dz_1 \int_0^\infty dz_2 \Psi^*_{Q+q/2}(z_1) \Psi_{Q-q/2}(z_2) \phi_d(z_1) \phi^*_d(z_2) \exp\left(j\kappa_z(z_1 - z_2)\right). \tag{4.117}$$

The integral $B(q, Q)$ determines the coupling between the waves of the discrete and continuous spectra due to the scattering on the inhomogeneities of $\delta\varepsilon$. The major contribution to $B(q, Q)$ comes from the spectral components of $\delta\varepsilon$ with the wavenumbers $|\kappa_z| > \mu = (-dU/dz)^{1/3}$ with $z < Z_s$ and the region of the altitudes $z_{1,2}$ in the neighbourhood of the turning point z_d of the mode of the discrete spectrum, where $U(z_d) = E_d$. We use the WKB approximation for the waves of the continuous spectrum and the Airy function representation for the waveguide mode in the vicinity of the turning point:

$$\Psi_E(z) = \sqrt{\frac{2}{\pi}}\,[U(z) - E]^{-1/4}\exp\left(-j\frac{3\pi}{4}\right)\sin\left(\int_0^z dz'\sqrt{U(z') - E}\right), \tag{4.118}$$

$$\phi_d(z) = \frac{1}{N_d}v(\mu(z - z_d)),\ N_d = \int_0^\infty dz\left|\phi_d(z)\right|^2.$$

The spatial spectrum of the fluctuations in the refractive index we define in the form (1.36) for the case of turbulent fluctuations $\delta\varepsilon$ isotropic in the (x, y)-plane

$$\Phi_\varepsilon(\vec{\kappa}) = 0.033C_\parallel^2\alpha\left(\kappa_\parallel^2 + \alpha^2\kappa_z^2\right)^{-11/6} \tag{4.119}$$

we substitute into Eqs. (4.117) and (4.118) the integral representation of the Airy function

$$v(t) = \frac{1}{2\sqrt{\pi}}\int_{-\infty}^{\infty} d\xi \exp\left(j\frac{\xi^3}{3} + j\xi t\right)$$

and require the inequalities

$$|\kappa_z| > \mu \text{ and } 2\sqrt{E_d}\frac{m}{k} >> 1 \tag{4.120}$$

to hold, which is necessary for the entire scheme of the solution obtained in Section 4.3. The meaning of inequality (4.120) is that we do not take into account the scattering of the leaky modes, assuming that the waveguide field is composed only of trapped (non-attenuated) modes. The $B(q, Q)$ can then be truncated to

$$B(q, Q) = C_\varepsilon^2\alpha^{-5/3}\times\begin{cases} 0.089\dfrac{\sin\left[q\Lambda_d(Q)\right]}{qN_d^2(E_d - Q)^{4/3}}, & \dfrac{\Delta\kappa}{\kappa_d} << 1 \\ 0.053\dfrac{\cos\left[q\Lambda_d(Q)\right]}{\mu^3N_d^2(E_d - Q)^{5/6}}, & \dfrac{\Delta\kappa}{\kappa_d} >> 1 \end{cases} \tag{4.121}$$

where

$$\Lambda_d = \frac{1}{2}\int_0^{z_d} dz'\sqrt{U(z') - Q},\quad \kappa_d = \sqrt{q_d - Q},\quad \Delta\kappa = \sqrt{U(0) - Q} - \sqrt{E_d - Q}. \tag{4.122}$$

Parameter κ_d determines the Bragg scattering angle θ_s from the waveguide mode into the wave of the continuous spectrum with energy Q at the height of the turning point $z = z_d$, i.e., $2\sin\theta_s/2 = \kappa_d/k$. The parameter $\Delta\kappa$ is a difference between the magnitude of the scattering angle at the surface $z = 0$ and at the altitude of the turning point $z = z_d$. Thus the inequality $\Delta\kappa/\kappa_d << 1$ requires that the change in the scattering angle owing to refraction in the scattering volume, i.e. in the region of the mode's localization, be much smaller than the scattering angle itself at the altitude z_d.

The relationship between the characteristic "energy" Q and the sliding angle θ of the wave front of the wave of the continuous spectrum relative to the surface $z = Z_s$ is given by the equation $Q = -k^2/m^2 \tan^2\theta$. The parameter $\Lambda_d(Q)$ determines the distance along x traversed by the wave of the continuous spectrum with "energy" Q as it propagates from $z = 0$ to $z = z_d$. Thus, $B(q, Q)$ defines the angular distribution of the scattered field, i.e., the scattering function in the "directions" E_1, E_2.

To calculate the intensity of the field $I_c(x, z)$ we employ the WKB approximation (4.118) and (4.121) in Eq. (4.116). We leave aside the constant terms for later and concentrate on the integration in Eq. (4.116). The integration over the "energy" difference q leads to the appearance of terms that place the limits on the region of the distances x' from which the scattered field is collected in the receiving antenna. With $\sqrt{U(0)} - \sqrt{q_d}/\sqrt{q_d} << 1$ we obtain

$$I_c(x,z) \sim \int_{-\infty}^{0} dQ\left(q_d - Q\right)^{-4/3}\left(U(z) - Q\right)^{-1/2}\int_0^x dx' \exp\left(\gamma_d x'\right)$$
$$\times\begin{Bmatrix} \frac{1}{2}\theta\left(\Lambda_d + \Lambda - \frac{x'}{k}\right) - \frac{1}{2}\theta\left(\Lambda_d - \Lambda - \frac{x'}{k}\right) - \\ \theta\left(\Lambda_d - \frac{x'}{k}\right)\cos\left(2\int_0^z dz'\sqrt{U\left(z'\right) - Q}\right) \end{Bmatrix} \tag{4.123}$$

and for $\frac{\sqrt{U(0)} - \sqrt{q_d}}{\sqrt{q_d}} >> 1$

$$I_c(x,z) \sim \int_{-\infty}^{0} dQ\left(q_d - Q\right)^{-5/6}\left(U(z) - Q\right)^{-1/2}\int_0^x dx' \exp\left(\gamma_d x'\right)$$
$$\times\begin{Bmatrix} \frac{1}{2}\delta\left(\frac{x'}{k} + \Lambda_d - \Lambda\right) + \frac{1}{2}\delta\left(\frac{x'}{k} - \Lambda_d + \Lambda\right) + \frac{1}{2}\delta\left(\frac{x'}{k} - \Lambda_d - \Lambda\right) \\ -\delta\left(\frac{x'}{k} - \Lambda_d\right)\cos\left(2\int_0^z dz'\sqrt{U\left(z'\right) - Q}\right) \end{Bmatrix} \tag{4.124}$$

where

$$\theta(x) = \begin{cases} 1, & x > 0 \\ 1/2, & x = 0 \\ 0, & x < 0 \end{cases}$$

and

$$\Lambda = \Lambda(Q, z) = \frac{1}{2}\int_0^z dz'\sqrt{U\left(z'\right) - Q}.$$

We introduce the dimensionless variables $\tau = -Qm^2/k^2$, $\tau_d = E_d m^2/k^2$, $h = kz/m$, $\beta = m^2\gamma_s/k$ and the modified profile $U(h) = m^2/k^2\,U(z)$. We then require that the inequality

$$k\gamma_d \Lambda_d(0) << 1 \tag{4.125}$$

holds. Since $\Lambda_d(0) \geq \Lambda_d(E_d)$, where $k\Lambda_d(E_d)$is the length of the cycle of a waveguide mode and the contributing values of Q are limited by the inequality $|Q| \leq E_d$, the condition (4.125) demands the smallness of the attenuation of the waveguide mode at the distance equal to its cycle. It may be noted that in the opposite case there is no waveguide mode structure to sustain scattering, the mode will be destroyed during one cycle.

Finally we obtain the intensity of the wave of the continuous spectrum normalized at the intensity of the free-space field

$$J_c(x,z) = I_c(x,z)\frac{x^2}{a^2} = 0.078k^{-5/3}xC_{\varepsilon\|}^2\alpha^{-5/3}m^{11/3}\left(\frac{|\phi_d(z_0)|^2}{N_d^2}\right)\times \\ \mathrm{K}(h)\exp\left(-\gamma_d x\right) \tag{4.126}$$

and the function $\mathrm{K}(h)$, which commands that the height distribution of the scattered field is given by two equations obtained from Eqs. (4.123) and (4.124).

With $\sqrt{U(0)} - \sqrt{\tau_d}/\sqrt{\tau_d} << 1$,

$$\mathrm{K}(h) = \int_0^\infty d\tau\left(\tau_d+\tau\right)^{-4/3}\left(U(h)+\tau\right)^{-1/2}\Lambda_d(\tau)\times \\ \left\{\exp\left(\beta\Lambda(\tau,h)\right) - \cos\left(2\int_0^h dh'\sqrt{U(h')+\tau}\right)\right\}, \tag{4.127}$$

and for $\sqrt{U(0)} - \sqrt{\tau_d}/\sqrt{\tau_d} >> 1$

$$\mathrm{K}(h) = \pi\left[-\frac{dU}{dh}\right]^{-2/3}\Bigg|_{h=h_d}\int_0^\infty d\tau\left(\tau_d+\tau\right)^{-5/6}\left(U(h)+\tau\right)^{-1/2}\times \\ \left\{\exp\left(\beta\Lambda(\tau,h)\right) - \cos\left(2\int_0^h dh'\sqrt{U(h')+\tau}\right)\right\}. \tag{4.128}$$

We shall now study the total field in the shadow region as being the result of composition of the waves of the discrete and continuous spectra excited due to the random scattering of the waveguide modes.

For the intensity of the waveguide mode the relation below follows:

$$J_d = \frac{x}{8\pi k|N_d|^4}|\phi_d(z_0)|^2|\phi_d(z)|^2\exp\left(-\gamma_d x\right). \tag{4.129}$$

We define the total intensity as

$$J_{tot} = J_d + J_c = \frac{x}{8\pi k \left|N_d\right|^2} \left|\phi_d(z_0)\right|^2 \exp\left(-\gamma_d x\right) S(z) \tag{4.130}$$

where

$$S(z) = \frac{\left|\phi_d(z)\right|^2}{N_d^2} + 0.139 k^{-2/3} C_\varepsilon^2 \alpha^{-5/3} m^{11/3} K(z). \tag{4.131}$$

The function $S(z)$ determines the height distribution of the total intensity of the field. We examine the behavior of the function $S(z)$ in the case of a bilinear dependence of $U(z)$. We assume the following values for the parameters of the problem: $k = 2\ \text{cm}^{-1}$, $a = 8500$ km, $Z_s = 11$ m, $\Delta\varepsilon_M = \varepsilon_M(0) - \varepsilon_M(Z_s) = 5.8\times 10^6$. The modified profile $U(h)$ can be defined as

$$U(h) = \begin{cases} \mu_1^3 (H_s - h), & h \geq H_s \\ h - H_s, & h > H_s \end{cases} \tag{4.132}$$

where $H_s = kZ_s/m = 2.42$, $\mu_1^3 = m^2 \Delta\varepsilon_M / H_s = 2$, $\tau_d = U(0) - \mu_1^2 \tau_1 = 1$, $U(0) = m^2 \Delta\varepsilon_M = 4.34$, $\tau_1 = 2.338$.

Consider the height dependence of the waveguide mode, i.e., the first term in $S(z)$. Thus far in the analysis of the waveguide field we have neglected the leakage of the trapped waves of the discrete spectrum through the potential barrier. The attenuation caused by the effect of the leakage is accounted for in the imaginary part of E_d, which can be written as

$$\delta_d = \operatorname{Im} E_d = \frac{\mu^2 \mu_1^2}{4\sqrt{\tau_1}} \exp\left\{ -\frac{4}{3} \tau_d^{3/2} \frac{1+\mu_1^3}{\mu_1^3} \right\}. \tag{4.133}$$

Sub-barrier leakage has virtually no effect on the height structure of the trapped mode inside the waveguide channel for $h < H_s$, where

$$\phi_d(h) = \frac{1}{N_d} v\left(\frac{\tau_d}{\mu_1^2} - \mu(H_s - h) \right). \tag{4.134}$$

Outside the waveguide channel the height structure of $\phi_d(h)$ corresponds to an outgoing wave with amplitude proportional to the leakage factor δ_d:

$$\phi_d(h) = \frac{\mu_1^{-1/3}}{\mu N_d} \tau_1^{1/4} \delta_d^{1/2} w_1 \left(\tau_d + j\frac{\delta_d}{\mu^2} - (h - H_s) \right). \tag{4.135}$$

Examining the second term in $S(z)$ given by Eq. (4.131), we can express the coefficient in front of $K(z)$ in terms of the non-dimensional attenuation coefficient β using Eq. (4.101) for γ_d from the previous section:

$$\beta = m^2 \gamma_d / k = 0.264 k^{-2/3} C_{\varepsilon\|}^2 \alpha^{-5/3} m^{11/3} \mu_1^{-5/9}.$$

Then for $S(z)$ we obtain

$$S(z) = \frac{\left|\phi_d(z)\right|^2}{N_d^2} + 0.492 \mu_1^{5/9} \beta K(z). \tag{4.136}$$

In principle, in order to determine the coefficient β it is necessary to know the magnitude of the structure constant C_ε and the anisotropy parameter of the irregularities of the refractive index in the volume of the waveguide channel. In some cases when the radio signal strength is measured, the measured data may be deployed to estimate the height dependence of the received signal. In particular, when the magnitude of the field attenuation per unit length $\gamma_x = dI/dx$ is known from the measured signal, then γ_d is related to γ_x via $\gamma_x = 4.34\gamma_d$, and given knowledge of the average M-profile we can construct a theoretical height structure of the field intensity $S(z)$. According to the data presented in Ref. [10], the γ_x magnitude for centimeter waves lies in the range $\gamma_x = 0.2$–0.5 dB km^{-1}. The corresponding limits for β are $\beta = 0.207$–0.52 for $k = 2$ cm^{-1} and $a = 8500$ km (normal refraction above the evaporation duct). Thus under real conditions the coefficient in front of $K(z)$ in Eq. (4.111) is of the order of unity and, consequently, the contribution of the waves of the continuous spectrum to $S(z)$ can be significant.

Figure 4.15 shows the result of the calculation of the resulting height distribution 10 log $S(h)$ based on Eq. (4.136) for $\beta = 0.207$ and $\beta = 0.52$ with $U(h)$ given by Eq. (4.132) and its parameters defined above. The figure also shows the height dependence of the first term in Eq. (4.136), i.e., the contribution of the waves of the discrete spectrum alone. As observed from the figure, the intensity of the field above the turning height h_d (in this case $h_d = 1.67$) is a contribution from the waves of the continuous spectrum, i.e. second term in Eq. (4.136). Apparently, scattering on ran-

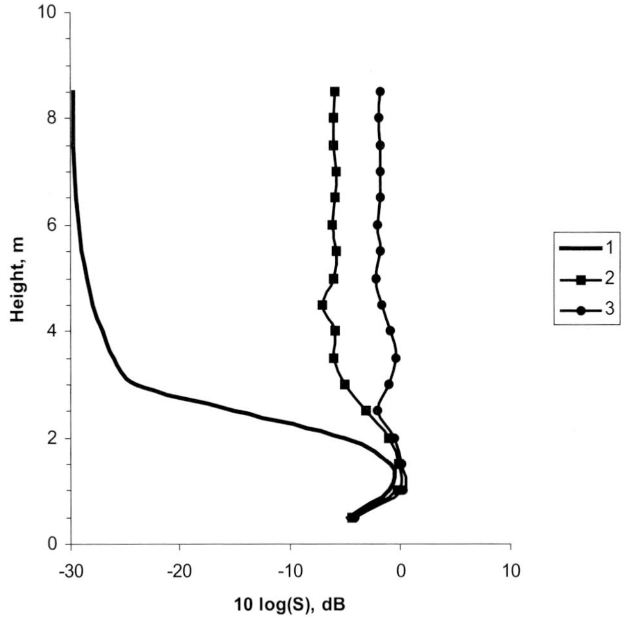

Figure 4.15 Height structure of the wave field in the presence of an evaporation duct: $f = 10$ GHz, Curve 1, no scattering on inhomogeneities of the refractive index, curves 2 and 3, scattering is taken into account.

dom fluctuations in the refractive index leads to a significant change in the height dependence of the signal strength; in the presence of random fluctuations $\delta\varepsilon$ in the refractive index there is no sharp exponential decay in the average signal strength outside the duct and for sufficiently strong fluctuations the $\delta\varepsilon$ field outside the duct may increase to the order of magnitude of the field inside, thus revealing a rather smooth unpronounced height dependence compared to the case of deterministic duct propagation.

At relatively large altitudes, $h - H_s >> \tau_d$, the function $S(h)$ has the following asymptotic form

$$S(h) \sim \frac{1}{4\sqrt{h - H_s - \tau_d}} \exp\left[-\frac{4}{3}\tau_d^{3/2}\frac{\mu_1^3 + 1}{\mu_1^3} + 2\frac{\delta_d}{\mu^2}\sqrt{h - H_s - \tau_d}\right]$$
$$+ \frac{1.58\mu_1^{5/9}\beta}{\tau_d^{1/3}\sqrt{h - H_s}} \exp\left(\beta\sqrt{h - H_s}\right). \tag{4.137}$$

Beginning with the altitudes $h_1 = H_s + \tau_d + \frac{\mu^4}{2\delta_d^2}$ and $h_2 = H_s + \frac{1}{4\beta^2}$, the exponential growth predominates over the cylindrical divergence in either the first or second terms, respectively. The exponential growth of those terms in Eq. (4.137) is due to the arrival of the waves from shorter distances $x' < x$, at which the field of the waveguide mode was exponentially large, compared with its value at distance x. The second term makes the main contribution in Eq. (4.137), since the exponential factor β exceeds the sub-barrier leakage factor δ by at least an order of magnitude.

As mentioned before, we ignored the scattering of the waves in the space above the waveguide. While, in principle, it should be taken into account for the sake of consistency, the qualitative character of the total field will not be expected to change drastically. We may assume however that additional scattering of the waves of the continuous spectrum in the space above the evaporation duct will smooth the exponential dependence of the scattered field in Eq. (4.137) for large altitudes. At the same time additional divergence of the waves due to a broadening of the angular spectrum can be compensated by the arrival of the scattered waves from the shadow region, i.e. from distances $x' < x - a/m\sqrt{h - H_s}$.

4.6 Evaporation Duct with Two Trapped Modes

Finally, we provide a closed form solution for the field intensity in the case when the evaporation duct can trap two modes. In this case the intensity in the two-mode waveguide is given by

$$J(x, z, z_0) = 4\pi x[I_1 \cdot S_1 + I_2 \cdot S_2] \tag{4.138}$$

where

$$I_1(x) = A_1 e^{\lambda_1 x} + A_2 e^{\lambda_2 x},$$

$$I_2(x) = A_3 e^{\lambda_1 x} + A_4 e^{\lambda_2 x},$$

$$\lambda_{1,2} = -\frac{\Gamma_1 + \Gamma_2}{2} \pm \frac{1}{2}\sqrt{(\Gamma_2 - \Gamma_1)^2 - 4W_{12}},$$

$$A_1 = -\frac{[I_{10}(\lambda_2 + \Gamma_2) - W_{12} I_{20}]}{\lambda_1 - \lambda_2},$$

$$A_2 = \frac{[I_{10}(\lambda_1 + \Gamma_1) - W_{12} I_{20}]}{\lambda_1 - \lambda_2},$$

$$A_3 = \frac{[I_{20}(\lambda_2 + \Gamma_2) - W_{12} I_{10}]}{\lambda_2 - \lambda_1},$$

$$A_4 = \frac{[I_{20}(\lambda_1 + \Gamma_1) - W_{12} I_{10}]}{\lambda_1 - \lambda_2},$$

$\Gamma_1 = 1.603 \cdot A_s$, $\Gamma_2 = 3.98 \cdot A_s$, $W_{12} = 0.316 \cdot A_s$, and $A_s = k^{-2/3} C_\varepsilon^2 \alpha^{-5/3} m^{11/3}$.

Coefficients $I_{n0} = |\chi_n(h_0)|^2$ are the coefficients of the excitation of the mode with number n and function $S_n(h)$ is a height distribution of the total intensity of the nth mode:

$$S_n(h) = |\chi_n(h)|^2 + 0.492\Gamma_n \mu_1^{5/3} P_n(h). \tag{4.139}$$

The height gain function χ_n is given by

$$\chi_n(h) = \frac{\sqrt{\mu_1}}{\tau_n} v\left(\frac{t_n - U(h)}{\mu_1^2}\right), \quad h < H_s \tag{4.140}$$

$$\chi_n(h) = \delta^{1/2} w_1\left(\frac{t_n - U(h)}{\mu_2^2}\right), \quad h \geq H_s$$

where δ is the coefficient of penetration through the potential barrier into the space above the duct:

$$\delta = \frac{\mu_1}{4\tau_n} \exp\left[-\frac{4}{3}\frac{\mu_2^3 + \mu_1^3}{\mu_1^3 \mu_2^3} t_n^{3/2}\right], \tag{4.141}$$

$$\tau_n = \left[\frac{3}{2}\pi\left(n - \frac{1}{4}\right)\right]^{2/3}.$$

The function $P_n(h)$ is given by the integral

$$P_n(h) = \pi/\mu_1^2 \int_0^\infty ds(t_n + s)^{-5/6} (U(s) + s)^{-1/2} \left[e^{\Lambda(s,h)} - \cos\left(2Q(s,h)\right) \right] \quad (4.142)$$

where

$$Q(s,h) = \frac{2}{3\mu_1^3} \left[(U_0 + s)^{3/2} - (U(h) + s)^{3/2} \right], \quad h < H_s \quad (4.143)$$

$$Q(s,h) = \frac{2}{3} \left[\frac{1}{\mu_1^3} (U_0 + s)^{3/2} - \frac{\mu_1^3 + \mu_2^3}{\mu_1^3 \mu_2^3} s^{3/2} + (U(h) + s)^{3/2} \right], \quad h \geq H_s$$

$$\Lambda(s,h) = \frac{1}{\mu_1^3} (U_0 + s)^{1/2} - \frac{1}{\mu_1^3} (U(h) + s)^{1/2}, \quad h < H_s \quad (4.144)$$

$$\Lambda(s,h) = \frac{1}{\mu_1^3} (U_0 + s)^{1/2} - \frac{\mu_1^3 + \mu_2^3}{\mu_1^3 \mu_2^3} s^{1/2} + (U(h) + s)^{1/2}, \quad h \geq H_s .$$

It should be noted that the above formulas actually represent the incoherent sum of two trapped modes. Under very moderate assumptions for the intensity of the atmospheric turbulent fluctuations (as in Figures 4.16 and 4.17 below) the interference term in the coherence function may be neglected since phase fluctuations in this case are strong enough to make the above approximation valid.

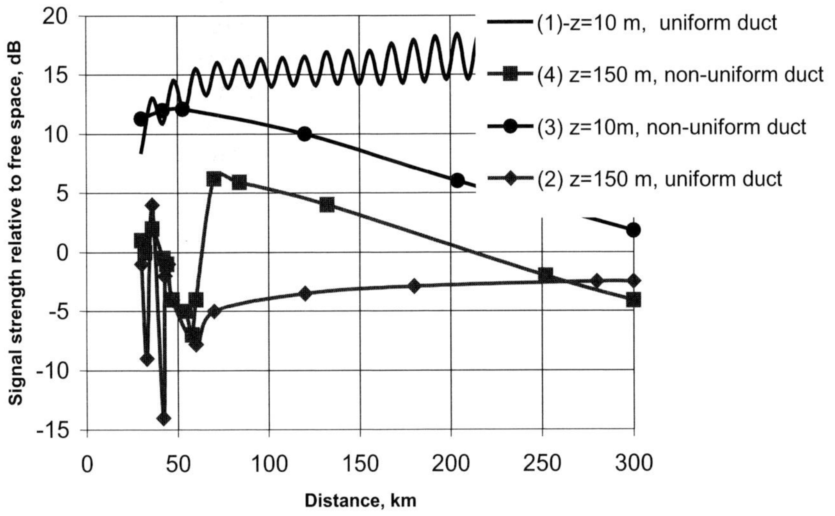

Figure 4.16 Range dependence of the signal strength at 10 GHz in the presence of an evaporation duct with two trapped modes. The transmitter antenna is mounted at 10 m above the sea surface. Curves (1) and (2) correspond to the range dependence of the received field strength for the receiving antennas at the heights inside (1) and outside (2) the duct for the case of a uniform duct in the absence of random fluctuations in the refractive index. Curves (3) and (4) correspond to the same antenna elevations for a non-uniform duct with random and anisotropic fluctuations of the refractive index: $C_\varepsilon^2 = 10^{-14}$ cm$^{-2/3}$, $\alpha = 0.1$.

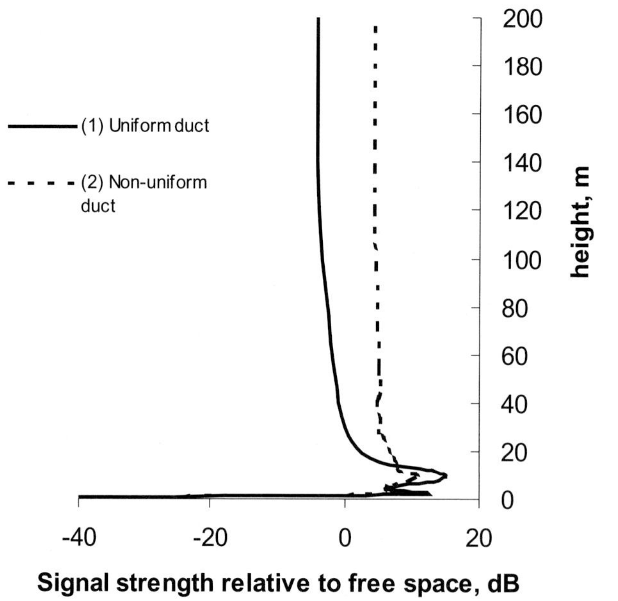

Figure 4.17 Height-gain dependence of the signal strength at 10 GHz in the presence of an evaporation duct with two trapped modes at 100 km from the transmitter. The transmitter antenna is mounted at 10 m above the sea surface. Curve (1) shows the received field versus the height in the absence of random fluctuations in the refractive index. Curve (2) is a height-gain dependence for a non-uniform duct with random and anisotropic fluctuations of the refractive index: $C_\varepsilon^2 = 10^{-14}$ cm$^{-2/3}$, $\alpha = 0.1$.

To illustrate the effect of scattering on random fluctuations in a two-mode evaporation duct we calculated the signal strength at frequency 10 GHz in the evaporation duct with parameters $Z_s = 15$ m and $\Delta M = 5$ N-units. Figure 4.16 shows a range dependence of the signal strength in the absence and presence of the fluctuations in the refractive index. As observed from the figure, at a height 10 m inside the duct the composition of the two trapped modes produces an interference pattern, creating fades. The second curve shows the range dependence at the height 150 m where the second trapped mode is dominant and therefore no interference pattern is observed. The height gain pattern in this case is typical for one that follows from the model of the evaporation duct, namely, revealing the maximum of the signal strength inside the duct, Figure 4.17. In the presence of fluctuations we observe significant attenuation of the field with distance as well as considerable changes in the height gain distribution, namely, the height structure becomes largely uniform.

In the range of frequencies from 3 GHz to 20 GHz the evaporation duct in most cases can support only a few trapped modes. The calculation of the field strength for a finite number of modes can be performed in a way similar to that described in this section. The surface-based and elevated duct created by an advection mechanism and normally associated with strong elevated inversion of temperature may trap a hundred modes and the above approach is inefficient. In that case other methods based on the application of diffusion theory for a parabolic equation can be employed to solve the problem of scattering in a multimode waveguide.

References

1 Bean, B.R., Cahoon, B.A., Samson, C.A. and Thayer, C.D. *A World Atlas of Atmospheric Radiorefractivity*, New York, 1966, 1113 pp.

2 Hitney, H.V., Richter, J.H., Pappert, R.A., Anderson, K.D. and Baumgartner, G.B. Tropospheric radio propagation assessment, 1985, *Proc. IEEE*, 1985, 73 (2), 265–283.

3 Gossard, E.E. Clear weather meteorological effects on propagation at frequencies above 4 GHz, *Radio Sci.*, 1981, 16 (5), 589–608.

4 Fock, V.A. Electromagnetic Diffraction and Propagation Problems, Pergamon Press, 1965.

5 Rotheram, S. Radiowave propagation in the evaporation duct, 1974, *The Marconi Rev.*, 1974, 67 (192), 18–40.

6 Booker, H.G. and Walkinshaw, W. The mode theory of tropospheric refraction and its relation to waveguides and diffraction, in *Meteorological Factors in Radio Wave Propagation*, Physics Society, London, 1946, pp. 80–127.

7 Bocharov, V.G., Kukushkin, A.V., Sinitsin, V.G. and Fuks, I.M. *Radio Propagation in Surface Tropospheric Ducts*, IRE, Ukrainian Acad. Sci., Preprint No 126, 1969, 44 pp.

8 Kerr, D.E. Transmission along the California coast, in *Propagation of Short Radio Waves*, Kerr, D.E. Ed., McGraw Hill, New York, 1951, pp. 328–335.

9 *Report on Factual Data from the Canterbury Project*, Department of Science and Industrial Research, Wellington, New Zealand, 1951.

10 Braude, S.Ya., Ostrovsky, I.E., Sanin, F.S. and Shamfarov, Ya.L. *Propagation of Centimetre Waves over the Sea in the Presence of Superrefractivity*, Morskoi Vestnik, Leningrad, 1949, No2, 103 pp.

11 Peceris, C.L. Wave theoretical interpretation of the propagation of the 10 cm waves in low-level ocean ducts, 1947, *Proc. IRE*, 1947, pp. 453–462.

12 Paulus, R.A. Practical application of an evaporation duct model, 1985, *Radio Sci.*, 1985, 20 (4), 887–896.

13 Heemskerk, H.J.M. Boekema, R.B. The influence of the evaporation duct on the propagation of electromagnetic waves low above the sea surface at 3–94 GHz, *Eighth International Conference on Antennas and Propagation*, 1993, Vol. 1, pp. 348–351.

14 Anderson, K.D. Radar measurements at 16.5 GHz in the oceanic evaporation duct, *IEEE Trans. Antennas Propagation*, 1989, 37 (1), 100–106.

15 Anderson, K.D. 94 GHz propagation in the evaporation duct, *IEEE Trans. Antennas Propagation*, 1990, 38 (5), 746–753.

16 Hitney, H.V., Vieth, R. Statistical assessment of evaporation duct propagation, *IEEE Trans. Antennas Propagation*, 1990, 38 (6), 794–799.

17 Hitney, H.V., Hitney, L.R. Frequency diversity effects of evaporation duct propagation, *IEEE Trans. Antennas Propagation*, 1990, 38 (10), 1694–1700.

18 Anderson, K.D. Radar detection of low altitude targets in a maritime environment, *IEEE Trans. Antennas Propagation*, 1995, 43 (6), 609–616.

19 Shen, X., Vilar, E. Path loss statistics and mechanisms of transhorizon propagation over a sea path, *Electron. Lett.*, 1996, 32 (3), 259–261.

20 Brookner, E., Ferraro, E. and Ouderkirk, G.D. Radar performance during propagation fades in the mid-Atlantic region, 1998, *IEEE Trans. Antennas Propagation*, 1998, 46 (7), 1056–1064.

21 Beckmann, P., Spizzichino, H. *The Scattering of Electromagnetic Waves from Rough Surfaces*, Pergamon and MacMillan, New York, 1963.

22 Baz, A.I., Zeldovich, Y.B., Perelomov, A.M. *Scattering, Reactions and Decays in Nonrelativistic Quantum Mechanics*, 1980, Academic Press, New York, 1980.

23 Brocks, K., Jeske, H. The meteorological conditions on electromagnetic wave propagation above the sea, in *Electromagnetic Distance*

Measurements, Pergamon Press, London, 1967, pp. 122–136.

24 Ott, R.H. Roots of the modal equation for em wave propagation in a tropospheric ducts, 1980, *J. Math. Phys.*, 1980, 21 (5), 1255–1266.

25 Belobrova, M.V., Ivanov, V.K., Kukushkin, A.V., Levin, M.B. and Fastovsky, J.A. Prediction system on UHF radio propagation conditions over the sea, Institute of Radio Astronomy, Ukrainian Academy of Science, Preprint No 31, 1989, 39 pp.

26 Kukushkin, A.V., Fastovsky, J.A. and Levin, M.B. Propagation effect analysis: program system, *Seventh International Conference On Antennas and Propagation*, Conference Publication No 333, Part 1, 1991, pp. 535–539.

27 Budden, K.C. *The Waveguide Mode Theory of Wave Propagation*, Pergamon Press, London, 1961.

28 Wait, J.R. *Electromagnetic Waves in Stratified Media*, Pergamon Press, New York, 1970.

29 Baumgartner, G.B., Hitney, H.V. and Pappert, R.A. Duct propagation modelling for the integrated refractive effects prediction system (IREPS), *Proc. IEE*, 1983, 130 Part F, 630–642.

30 Marcus, S.W. A model to calculate em fields in tropospheric duct environment at frequencies through SHF, *Radio Sci.*, 1982, 17, 895–901.

31 Tatarskii, V.I. *The Effects of Turbulent Atmosphere on Wave Propagation*, IPST, Jerusalem, 1971.

32 Kukushkin, A.V., Freilikher, V.D. and Fuks, I.M. Over-the-horizon propagation of UHF radio waves above the sea , *Radiophys. Quantum Electron.* (transl. from Russian), Consultant Bureau, New York , *RPQEAC*, 1987, 30 (7), 597–620.

5
Impact of Elevated M-inversions on the UHF/EHF Field Propagation beyond the Horizon

The elevated refractive layer, the so-called M-inversion, is frequently associated with an anomalousy high level of the received signal at UHF frequencies over the horizon. It is apparent that the methods of prediction of either the parameters of the elevated layer or the signal level are needed in many applications, such as radar, surveillance and communications.

The methods of the analytical solutions to the problem of wave propagation in the presence of elevated M-inversion are less well developed than those applied to propagation in an evaporation duct. The major reason for this is that such a waveguide has a multi-mode or multi-ray nature that, in turn, may require a different approach to obtaining the analytical solution, depending on the geometry of the problem (positions of the transmitter and receiver relative to the elevated duct "boundaries", range and frequency).

For instance, when the transmitter and receiver are based close to the ground with the distance between them $x << 2\sqrt{2aZ_i}$, see Figure 5.1, where Z_i is the height of the minimum in the M-profile of the refractivity, the most effective approach is to apply the method of multiple reflections [1, 2], that, in turn leads to the approximation of the geometrical optic with $k \to \infty$. In contrast, at longer distances, the most effective method uses normal waves providing the modal representation of the wave field.

In a sub-tropical region of the world's oceans both evaporated and elevated ducts may exist simultaneously thus complicating the situation. In Ref. [3] this situation is analysed by applying the normal wave method to the M-profile approximated by a piece-wise linear profile. An alternative approach is described in Section 5.2, where the contribution of evaporation duct is presented by trapped modes while the reflection from elevated M-inversion is analysed in terms of geometric optics.

This chapter is arranged as follows. The modal representation of the wave field for the case of elevated M-inversion is presented in Section 5.1. Section 5.2 introduces the hybrid, ray and modes, representation of the wave field in the problem of a two-channel system. Some results of the measurements versus prediction are discussed in Section 5.3 and, finally, in Section 5.4, we introduce a method of estimating the excitation of the elevated duct due to scattering of the direct wave on the fluctuations in refractivity in the vicinity of the upper boundary of the atmospheric boundary layer.

Radio Wave. Alexander Kukushkin

ISBN: 3-527-40458-9

5.1
Modal Representation of the Wave Field for the Case of Elevated M-inversion

Consider a piece-wise linear model of the M-profile as shown in Figure 5.1 and introduce the dimensionless coordinates $\xi = mx/a$, $h = kz/m$. The parameters of the M-profile can be expressed in terms of h-coordinates and the scaled potential $U(h) = 2m^2 10^{-6} M(z)$:

$$H_k = kZ_k/m, \quad H_i = kZ_i/m, \quad H_s = kZ_s/m,$$
$$U_k = 2m^2 10^{-6} M(Z_k), \quad U_i = 2m^2 10^{-6} M(Z_i), \quad U_0 = 2m^2 10^{-6} \Delta M.$$

Let us also introduce the gradients of the M-profile in each of the layers between Z_s, Z_i and above Z_k respectively:

$$G_2 = dM/dz, \quad \text{with } Z_s < z < Z_i$$
$$G_4 = dM/dz, \quad \text{with } z > Z_k.$$

The respective gradients of the dimensionless profile $U(h)$ are given by

$$\mu_1^3 = U_0/H_s, \mu_2^3 = a \cdot 10^{-6} G_2, \mu_3^3 = (U_i - U_k)/(H_k - H_i), \mu_4^3 = a \cdot 10^{-6} G_4.$$

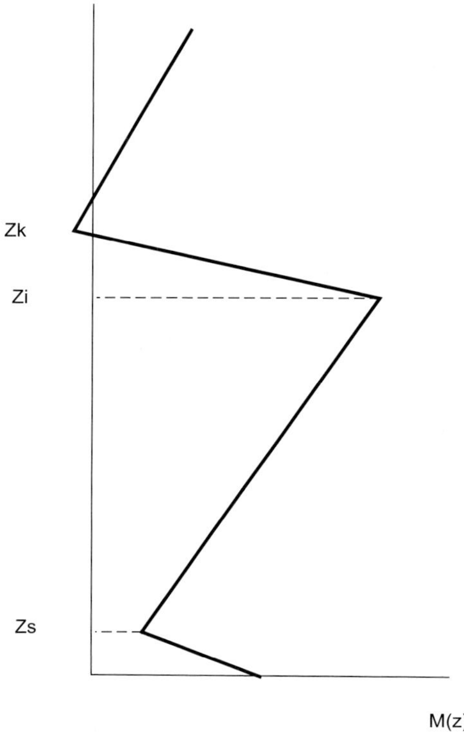

Figure 5.1 Piece-wise linear approximation of the M-profile with two M-inversions.

Following the approach described in Chapter 2, we expand the attenuation factor $W(\xi, h, h_0)$ over the set of eigenfunctions of a continuum spectrum. Using the results of Section 2.3 and introducing the dimensionless "energy" $t = m^2 E/k^2$ we obtain

$$W(\xi, h, h_0) = e^{-j\frac{\pi}{4}} \int_{-\infty}^{\infty} dt \Psi(t, h) \Psi^*(t, h_0) e^{j\xi t} \tag{5.1}$$

where the eigenfunction $\Psi(t, h)$ obeys the equation

$$\frac{d^2\Psi}{dh^2} + [U(h) - t]\Psi = 0 \tag{5.2}$$

and the boundary conditions

$$\Psi(t, 0) = 0 \text{ and } \Psi(t, \infty) = 0\,. \tag{5.3}$$

As known from analogy with a quantum-mechanical problem [4], the solution to Eq. (5.2) is given by a superposition of the waves, outgoing (χ^+) to infinity ($h \to \infty$) and incoming (χ^-) from infinity:

$$\Psi(t, h) = \frac{1}{2\sqrt{\pi\mu_4}} \left[\chi^-(t, h) - S(t)\chi^+(t, h)\right]. \tag{5.4}$$

It is also observed that $\chi^-(t, h) = \left(\chi^+(t, h)\right)^*$, where the sign * indicates a complex conjugate. The factor $\frac{1}{2\sqrt{\pi\mu_4}}$ in Eq. (5.4) comes from a delta-function normalisation of the eigenfunction of the continuum spectrum, Section 2.3. From the boundary condition (5.3) we have

$$S(t) = \frac{\chi^-(t,0)}{\chi^+(t,0)}. \tag{5.5}$$

Given the linear approximation to the M-profile in each of the layers: $H_s < h \le H_i$ $H_s < h \le H_i$, $H_i < h \le H_k$ and $h > H_{ki}$ the solution for $\chi^\pm$ can be presented via superposition of the Airy–Fock function. Taking into account the continuity of both the function $\chi^\pm$ and its derivative at the boundaries of the layers with constant gradient of $U(h)$, the outgoing wave can be written as follows:

1. With $h > H_k$

$$\chi^+(h) = w_1\left(\frac{t - U(h)}{\mu_4^2}\right). \tag{5.6}$$

2. With $H_k \ge h > H_i$

$$\chi^+(h) = A\left\{w_1\left(\frac{t - U(h)}{\mu_3^2}\right) + R_k w_2\left(\frac{t - U(h)}{\mu_3^2}\right)\right\} \tag{5.7}$$

where

$$A = \frac{w_2(x_k^+)}{w_1(x_k^-) + R_k w_2(x_k^-)}, \tag{5.8}$$

$$R_k = -\frac{w_2'(x_k^-) + Q_k w_2(x_k^-)}{w_1'(x_k^-) + Q_k w_1(x_k^-)}. \tag{5.9}$$

Here $x_k^+ = \frac{t - U_k}{\mu_4^2}$, $x_k^- = \frac{t - U_k}{\mu_3^2}$, R_k is a reflection coefficient of the wave incident to the boundary $h = H_k$ from $h < H_k$, the parameter Q_k is the surface impedance at the boundary $h = H_k$ and is given by

$$Q_k = \frac{\mu_4}{\mu_3} \frac{w_1'(x_k^+)}{w_1'(x_k^-)}. \tag{5.10}$$

3. With $H_i \geq h > H_s$

$$\chi^+(h) = B\left\{ w_1\left(\frac{t - U(h)}{\mu_2^2}\right) + R_i w_2\left(\frac{t - U(h)}{\mu_2^2}\right)\right\} \tag{5.11}$$

where

$$B = A\frac{w_2(x_i^+) + R_k w_1(x_i^+)}{w_1(x_i^-) + R_i w_2(x_i^-)}, \tag{5.12}$$

$$R_i = -\frac{w_1'(x_i^-) + Q_i w_1(x_i^-)}{w_2'(x_i^-) + Q_k w_2(x_i^-)}, \tag{5.13}$$

$$Q_i = \frac{\mu_3}{\mu_2} \frac{w_1'(x_i^+) + R_k w_1(x_i^+)}{w_2'(x_i^-) + R_k w_2(x_i^-)}. \tag{5.14}$$

Here $x_i^+ = t - U_i/\mu_3^2$, $x_i^- = t - U_i/\mu_2^2$, the other parameters have the same meaning as above.

4. With $h \leq H_s$

$$\chi^+(h) = C\left\{ w_2\left(\frac{t - U(h)}{\mu_1^2}\right) + R_s w_1\left(\frac{t - U(h)}{\mu_1^2}\right)\right\} \tag{5.15}$$

where

$$C = B\frac{w_1(x_s^+) + R_i w_2(x_s^+)}{w_2(x_s^-) + R_s w_1(x_s^-)}, \tag{5.16}$$

$$R_s = -\frac{w_2'(x_s^-) + Q_s w_2(x_s^-)}{w_1'(x_s^-) + Q_s w_1(x_s^-)}, \tag{5.17}$$

$$Q_i = -\frac{\mu_2}{\mu_1} \frac{w_1'(x_s^+) + R_i w_1(x_s^+)}{w_2'(x_i^-) + R_k w_2(x_i^-)}. \tag{5.18}$$

Here $x_s^+ = \tau/\mu_2^2$, $x_s^- = \tau/\mu_1^2$, the other parameters have the same meaning as above.

The solution to $W(\xi, h, h_0)$ is obtained in principle and given by Eqs. (5.1)–(5.18). However, the direct calculation of Eq. (5.1) is not practical because of the problem of convergence. Therefore, a significant amount of research, see for example Refs. [3, 5, 6–10] has been dedicated to finding effective methods of calculating the electromagnetic field in the presence of elevated M-inversion.

Here we consider a modal representation of the attenuation function $W(\xi, h, h_0)$ in a two-channel system when both elevated and surface-based M-inversion are present. In this approach, the integral (4.1) is calculated as a sum of the residue at the poles of the S-matrix in the upper half-space of variable t. Taking into account Eqs. (5.5) – (5.18) we obtain the expression for the S-matrix in this particular problem:

$$S = -R_g \frac{A^* w_2^2(x_s^-)\cdot(1-R_i^* T_s^-)\cdot(1-R_i^* T_s^+)\cdot(1-R_i^* R_g^*)}{A w_1^2(x_s^-)\cdot(1-R_i T_s^+)\cdot(1-R_i T_s^- *)\cdot(1-R_i R_g)} \tag{5.19}$$

where $R_g = -w_1(x_0)/w_2(x_0)$ is the reflection coefficient of the ground, $x_0 = t - U_0/\mu_1^2$, $T_s^+ = -w_2(x_s^+)/w_1(x_s^-)$ is the reflection coefficient of the wave incoming from the upper space from the boundary $h = H_s$with ideal boundary conditions on it; $T_s^- = -w_1(x_s^-)/w_2(x_s^+)$ is a similar coefficient of reflection from the boundary $h = H_s$,but for the wave incoming from the space below the boundary $h = H_s$.

The resonant terms in the denominator of the S-matrix determine the spectrum of the normal waves in a two-channel system. Apparently, the propagation constants of the normal waves are defined by the position of the poles of the S-matrix in the t-plane. The waves trapped in an evaporation duct (surface-based M-inversion) have propagation constants t_n in the interval Re $t_n \in (0, U_0)$. Waves localised in the elevated duct between H_s and H_k have propagation constants t_n in the interval Re $t_n \in (0, U_i)$. And, finally, the normal waves localised in a channel formed by the ground surface $h = 0$ and the upper boundary of the elevated layer $h = H_k$ are in the interval Re $t_n \in (U_k, 0)$.

Now we consider the resonant terms of the S-matrix in more detail and assume that $U_k > 0$. The resonant term $1 - R_i T_s^+$ determines the spectrum of the normal waves for the elevated channel with boundaries $h = H_s$ and $h = H_k$. With $U_0 \neq 0$ localisation of the modes in the surface-based channel is possible, in principle, and the spectrum of these modes is defined by the equation $1 - R_s R_g = 0$. In the case where $U_k U_0$, both series of propagation constants are clearly separated in the wave-number space. More detailed analysis of this case is provided in Section 5.2. Now we concentrate on the modes of the elevated channel (duct). The characteristic equation for the spectrum of propagation constants is given by

$$1 - R_i T_s^+ = 0\,. \tag{5.20}$$

With $\tau/\mu_2^2 >> 1$ we can estimate $T_s^+ = -1 + O(e^{-\frac{4}{3} t_n^{3/2}/\mu_2^2})$ and Eq. (5.20) can then be reduced to $1 + R_i = 0$. Assume that the propagation constants of interests are such that Re $t_n - U_k/\mu_2^2 >> 1$. These normal waves experience complete reflection from the boundary $h = H_k$, in fact the respective rays turn back long before reaching the boundary H_k. The reflection coefficient can then be estimated as $R_k \approx -1$. Taking into account Eq. (5.12) for R_i we can obtain instead of Eq. (5.20) the following characteristic equation for the waves trapped in an elevated duct

$$v'(x_i^-) + q_i v(x_i^-) = 0 \tag{5.21}$$

where $q_i \cong \mu_3/\mu_2\ v'(x_i^+)/v(x_i^+)$, $v(x) = (1/2)j[w_1(x) - w_2(x)]$. Equation (5.21) can be further simplified using the asymptotic formulas for the Airy function v. The result is given by

$$\sin\left(\varsigma_i^+ + \varsigma_i^- + \frac{\pi}{2}\right) = 0, \tag{5.22}$$

where

$$\varsigma_i^+ = \frac{2}{3\mu_3^3}(U_i - t)^{3/2}, \quad \varsigma_i^- = \frac{2}{3}\mu_2^3(U_i - t)^{3/2}.$$

The solution to Eq. (5.22) provides a spectrum of the propagation constants t_n of the elevated duct:

$$t_n = U_i - \left[\frac{3}{2}\frac{(\mu_2\mu_3)^3}{\mu_2^3+\mu_3^3}\pi\left(n - \frac{1}{2}\right)\right]^{2/3}. \tag{5.23}$$

As observed, Eq. (5.23) is obtained by neglecting the leaking of the modes into the space outside the duct. The number N_1 of trapped modes is limited by the condition $t_n > U_k$, therefore,

$$N_1 = \text{entier}\left[\frac{2}{3}\frac{\mu_2^3+\mu_3^3}{(\mu_2\mu_3)^3}(U_i - U_k)^{3/2} + \frac{1}{2}\right]. \tag{5.24}$$

Counting the next terms of the asymptotic for both T_s^+ and R_k we can obtain from Eq. (5.20) some correction terms to the propagation constants t_n (5.23). These correction terms provide an estimate for a phase shift and attenuation factor for the modes due to the limited thickness of the potential barrier. Omitting the simple but cumbersome calculations we provide the final result for the imaginary part of the propagation constants:

$$\gamma_n = \text{Im}\ t_n = \frac{1}{4}\frac{\mu_2^3+\mu_3^3}{(\mu_2\mu_3)^3}\exp\left\{-\frac{4}{3}\frac{\mu_3^3+\mu_4^3}{(\mu_3\mu_4)^3}(t_n - U_k)^{3/2}\right\}. \tag{5.25}$$

Let us consider another situation and assume now that $U_k < 0$. We pay attention to the modes of the channels formed by the earth's surface $h = 0$ and the upper boundary of the M-inversion, $h = H_k$. As discussed above, the propagation constants of these modes lie in the interval $\mathrm{Re}\, t_n \in (U_k, 0)$. The spectrum of the propagation constant is determined by the characteristic equation $1 - R_s R_g = 0$. Using asymptotic expressions for Airy functions we obtain

$$R_s \cong R_i D e^{-j\delta}, \tag{5.26}$$

$$R_g \cong -\exp\left[j\frac{\pi}{2} + j\frac{4}{3\mu_1^3}(U_0 - t)^{3/2}\right]$$

where

$$\delta = \pi + \frac{4}{3}\frac{\mu_2^3 + \mu_3^3}{\mu_2^3 \mu_3^3}(-t)^{3/2}, \quad D = \frac{1}{2 - R_i T_s^+},$$

$$T_s^+ \cong -\exp\left[j\frac{\pi}{2} - j\frac{4}{3\mu_2^3}(-t)^{3/2}\right].$$

Let us expand the term D into a series

$$D = \sum_{m=0}^{\infty} (-1)^m (1 - R_i T_s^+)^m. \tag{5.27}$$

The series (5.27) takes into account the multiple effect of secondary reflection of the waves in the channel $H_s \le h \le H_k$ due to a leakage of the energy of the normal waves localised in the channel $0 < h \le H_k$ due to partial reflection from the boundary $h = H_s$. In the first and rough approximation this effect of mutual coupling of two channels can be neglected, at least this approximation will be good enough for $|\mathrm{Re}\, t_n| >> 1$. Under this condition we can retain only the first term in series (5.27). As a result we obtain

$$1 - R_s R_g \approx 1 - R_i R_g e^{-j\delta} = 0. \tag{5.28}$$

The last equation can be further simplified to the form

$$\frac{2}{3}\frac{\mu_2^3 + \mu_3^3}{\mu_2^3 \mu_3^3}(U_i - t_n)^{3/2} + \frac{2}{3\mu_1^3}(U_0 - t_n)^{3/2} - \frac{2}{3}\frac{\mu_2^3 + \mu_1^3}{\mu_2^3 \mu_1^3}(-t_n)^{3/2} = \pi\left(n - \frac{1}{4}\right). \tag{5.29}$$

The limiting case when surface M-inversion is absent is accounted for by Eq. (5.29) if we assume that $U_0 = 0$. We can also observe that the number n satisfying Eq. (5.29) starts from $n = N_1 + N_2 + 1$, where N_2 is the number of modes trapped in a surface-based channel, $0 < h \le H_s$.

To conclude, we may state that the characteristic equations (5.20) and (5.29) determine the limited, yet large, (for high frequencies and strong inversions of temperature) set of trapped modes in a two-channel system. While in the general case the two modal series in the evaporation and the elevated duct are coupled, for modes localised deep in respective channels the mutual coupling can be neglected in the first approximation. In this way, the trapped modes of the evaporation duct can be estimated as the modes formed by a surface-based inversion only, Figure 5.2, and the analysis is similar to that provided in Section 4.1.

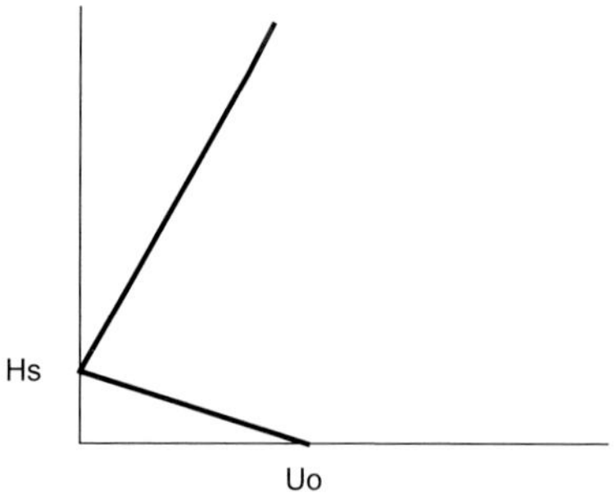

Figure 5.2 Isolated surface-based M-inversion.

Let us obtain a residue of the integrand in Eq. (5.1) in the pole of the S-matrix. Utilising the asymptotic expression for Airy functions incorporated into $\Psi(t, h)$ and $\Psi^*(t, h_0)$, the residue with Re $t_n > U_k$ is truncated to

$$\text{Re s}\,\Psi(t, h)\Psi^*(t, h_0)\big|_{t=t_n} = 2j\,\text{Im}\{t_n\}\overset{+}{\chi}(t_n, h)\overset{+}{\chi}(t_n, h_0) \tag{5.30}$$

Next we obtain an asymptotic expression for the height-gain functions $\overset{+}{\chi}(t_n, h)$. First, define the coefficients (5.8) and (5.12) for Re $t_n > U_k$:

$$A(t_n) \approx j\sqrt{\frac{\mu_4}{\mu_3}}\exp\left(\frac{\Delta}{2}\right) \tag{5.31}$$

where

$$\Delta = \frac{4}{3}\frac{\mu_3^3 + \mu_4^3}{\mu_3^3\mu_4^3}(t_n - U_k)^{3/2}$$

and, for Re $t_n > 0$,

$$B(t_n) \approx A\sqrt{\frac{\mu_3}{\mu_2}}\frac{1}{\cos(n\pi)}. \tag{5.32}$$

For $0 >$ Re $t_n > U_k$,

$$B(t_n) \approx 2j\sqrt{\frac{\mu_3}{\mu_2}}\sin\left(\frac{3}{2\mu_2^3}(U_i - t_n)^{3/2} + \frac{\pi}{4}\right)\left[1 - \exp\left\{j\frac{4}{3\mu_2^3}(U_i - t_n)^{3/2} + j\frac{\pi}{2}\right\}\right] \times \exp\left\{-j\frac{2}{3\mu_2^3}(U_i - t_n)^{3/2} - j\frac{\pi}{4}\right\}. \tag{5.33}$$

While obtaining Eq. (5.32) we take into consideration that $R_i(t_n) = -1$ at the pole. Assume that there is no evaporation duct present, i.e. $H_s = 0$, $U_0 = 0$, and consider a height-gain function $\overset{-}{\chi}(t_n, h)$ with $h \to 0$ for $0 >$ Re $t_n > U_k$. As follows from Eq. (5.11),

$$\overset{+}{\chi}(t_n, h = 0) = 2B\cos(S_1 - \delta)\exp(j\delta) \tag{5.34}$$

where

$$S_1 = \frac{2}{3\mu_2^3}(-t_n)^{3/2} + \frac{\pi}{4}, \; \delta = \frac{2}{3}\frac{\mu_2^3 + \mu_3^3}{\mu_2^3 \mu_3^3}(U_i - t_n)^{3/2}.$$

Taking into account that at the pole, as follows from Eq. (5.22) with $U_0 = 0$, the arguments of the cosine in the term $S_1 - \delta = -n\pi + \pi/2$, we obtain the boundary condition of interest, $\chi^+(h = 0) = 0$. With $U_0 \neq 0$ and Re $t_n < 0$, from Eq. (5.15) it follows that

$$\chi^+(t_n, 0) \approx \cos\left[\frac{2}{3\mu_1^3}(U_0 - t_n)^{3/2} - \frac{2}{3}\frac{\mu_1^3 + \mu_2^3}{\mu_1^3 \mu_2^3}(-t_n)^{3/2} - \frac{\pi}{4}\right]. \tag{5.35}$$

The equation for the poles takes the form $R_s R_g = 1$ in this case. The relationship between R_s and R_g is given by Eq. (5.26) from which we obtain that the argument of the cosine in Eq.(5.35) has the value $n\pi + \pi/2$ at the pole and, therefore $\chi^+(t_n, 0) = 0$.

In the case of positive Re $t_n > 0$, $|t_n| >> 1$, the behavior of the height-gain function $\chi^+(t_n, h)$ with $h \to 0$ is governed by the exponent factor

$$\chi^+(t_n, h)\Big|_{h\to 0} \approx \exp\left(-\frac{2}{3\mu_2^3}(-t_n)^{3/2}\right) \to 0$$

and the boundary condition at $h = 0$ is satisfied asymptotically.

Finally, we can present an explicit expression for the attenuation function $W(\xi, h, h_0)$ over the sum of the normal waves that can be used for computer calculation:

$$W(\xi, h, h_0) = e^{j\pi/4}\sqrt{\frac{\xi}{\pi}\frac{1}{4\mu_4}\frac{\mu_2^3 \mu_3^3}{\mu_2^3 + \mu_3^3}} \sum_{n=1}^{N} e^{j\xi t_n - \gamma_n \xi} \chi_n(h)\chi_n(h_0). \tag{5.36}$$

The number of trapped modes N is determined by the number of real roots of Eqs. (5.22) and (5.29). The eigenfunction $\chi_n(h)$ is given by the following equations:

1. With $h > H_k$

$$\chi_n(h) = w_1\left(\frac{t_n - U(h)}{\mu_4^2}\right)\exp\left(-\frac{\Delta}{2}\right). \tag{5.37}$$

2. With $H_k > h \geq H_i$

$$\chi_n(h) = 2\sqrt{\frac{\mu_4}{\mu_3}}\, v\left(\frac{t_n - U(h)}{\mu_3^2}\right). \tag{5.38}$$

3. With $H_i > h, \quad t_n > 0$

$$\chi_n(h) = 2jB(t_n)\, v\left(\frac{t_n - U(h)}{\mu_2^2}\right) \tag{5.39}$$

and for $H_i > h \geq H_s, \quad t_n < 0$

$$\chi_n(h) = B(t_n)\left[w_1\left(\frac{t_n - U(h)}{\mu_2^2}\right) + R_i(t_n) w_2\left(\frac{t_n - U(h)}{\mu_2^2}\right)\right]. \tag{5.40}$$

The factor $B(t_n)$ is defined by Eqs. (5.32) and (5.33).

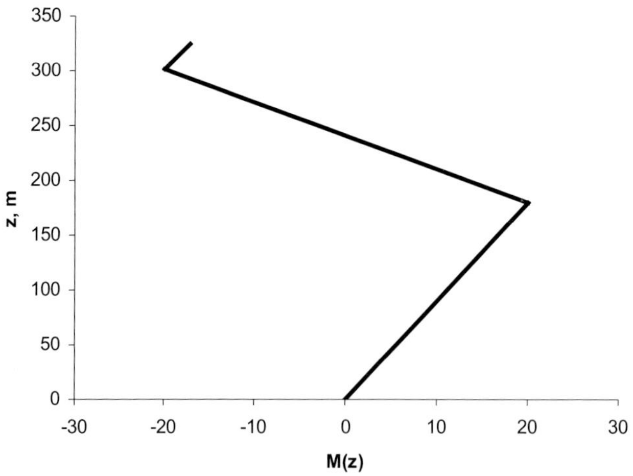

Figure 5.3 M-profile from Ref. [10].

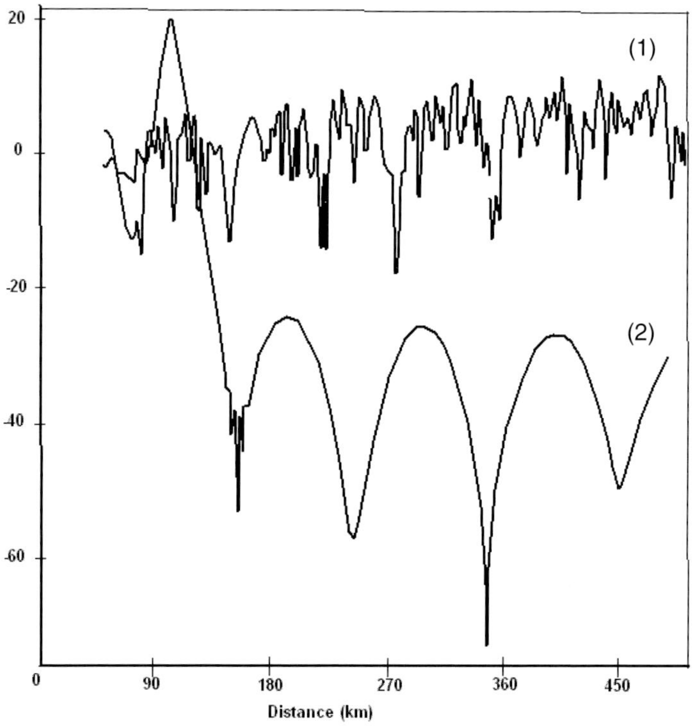

Figure 5.4 Range dependence of the received field at the wavelength 9.1 cm inside the duct (1) and above the duct (2).

4. With $H_s > h, \quad t_n < 0$

$$\chi_n(h) = \sqrt{\mu_1 \varsigma_n}^{-1/4} v(\mu_1 (h - \varsigma_n)), \tag{5.41}$$

where $\varsigma_n = \mu_1 H_s - t_n/\mu_1^2$.

The representation (5.36) is valid in a "geometric shadow" region, i.e. at a distance exceeding the maximum length of the cycle of the trapped modes $\Lambda_{\max} \sim 2\sqrt{2aZ_i}$. In the opposite case, at distances less than $\Lambda_{\max}$, the "leak" modes of a higher order provide a substantial contribution to the sum of the normal waves. These modes, though attenuating with distance, are not localised in the elevated duct, their amplitude grows exponentially with the order of the mode. The sum of leaked modes converges slowly as a result of interference of these modes. This leads to a need for precise determination of the relative phases of the modes and, in turn, to a sophisticated calculation of the complex propagation constants [5]. An alternative method to calculate the field at the "line-of-sight" distance $x < \Lambda_{\max}$ is to use the method of stationary phase applied directly to integral (5.1). In a transition region at distances

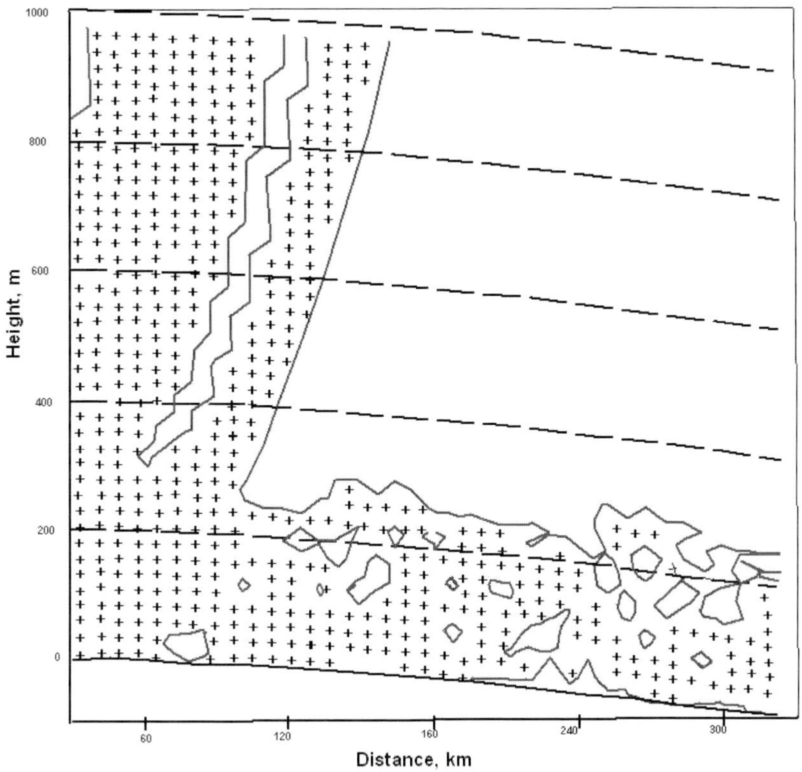

Figure 5.5 Coverage diagram for 10 GHz source at 30.5 m for the surface-based duct formed by M-profile in Figure 5.3. The return path loss is 220 dB.

close to Λ_{max}, the most practical approach is to use interpolation over values of the attenuation function obtained in either region. While the asymptotic expression can be obtained for the transition region it is very cumbersome and practically does not provide significant advantage over a simple interpolation.

To illustrate the modal representation described in this section, we apply the above formulas to the case of radio wave propagation at wavelength $\lambda = 9.1$ cm in a surface based tropospheric duct created by M-profile, shown in Figure 5.3. In fact this profile corresponds to the conditions of the experiment reported in Ref. [10]. Figure 5.4 shows the range dependence of the received signal strength at two elevations: 100 and 500 m. The calculated signal strength clearly illustrates the multimode character of the field inside the duct at the height 100 m and the contribution of a few modes only for the receiving antenna above the duct (at 500 m). Figure 5.5 shows the impact on the radar coverage diagram at 10 GHz in case of the above tropospheric duct. It is observed that strong ducting mechanism allows for a target detection at the distances of several hundred miles.

5.2
Hybrid Representation

In this section we use a Fock's contour integral representation for the attenuation factor $W(\xi, h, h_0)$. As shown in Chapter 2, both the contour integral representation and the expansion over the set of eigenfunctions of the continuous spectrum are equivalent in representing the attenuation factor $W(\xi, h, h_0)$ in the absence of random inhomogeneities of the refractive index. From practical point of view, in a deterministic problem the manipulations with a contour integral are a bit less cumbersome than eigenfunctions of the continuous spectrum, since the latter contains terms, such as waves $\chi^{-}(t, h)$ coming from infinity that produce a negligible contribution to the integral for positive ξ in the absence of random fluctuations.

In line with the study in Ref. [11], we consider a slightly more general case of the M-profile for the hybrid ray-mode representation, as shown in Figure 5.6.

The contour integral for the attenuation factor $W(\xi, h, h_0)$can be written as

$$W(\xi, h, h_0) = \sqrt{\frac{\xi}{\pi}} e^{-j\pi/4} \int_C K(h, h_0, t) e^{j\xi t}\, dt. \tag{5.42}$$

The function $K(h, h_0, t)$ satisfies the equation

$$\frac{d^2}{dh^2} K + [U(h) - t] K = \delta(h - h_0) \tag{5.43}$$

and the conditions

$$\left(\frac{dK}{dh} + qK\right)_{h=0} = 0, \tag{5.44}$$

$$\frac{d}{dh} \arg K \bigg|_{h\to\infty} > 0,$$

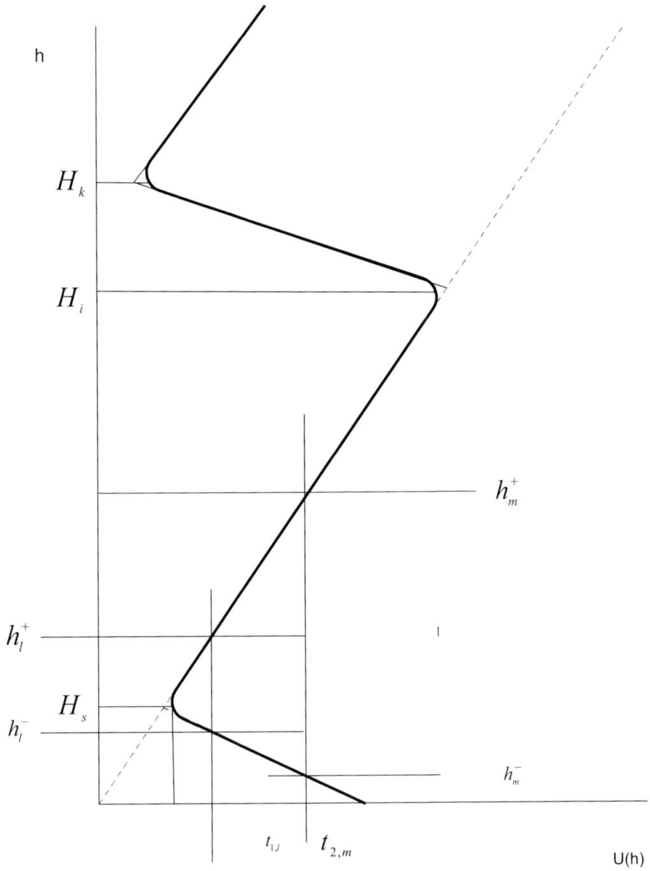

Figure 5.6 Smooth M-profile for a two-channel model of refractivity.

where $q = jmZ$ and Z is a surface impedance of the interface $h = 0$. The integration contour embraces singularities of the integrand lying in the first quadrant. The precise form of K depends on the shape of the profile $U(h)$ and the relative values h, h_0, H_s and H_i. If both the transmitter and the receiver are within the lower inversion layer, i.e. $h_0 < h < H_s$, then K takes the form

$$K(h, h_0, t) = -\frac{[f_2(h_0,t)+R_g f_1(h_0,t)]\cdot[f_1(h,t)+R\cdot f_2(h,t)]}{W(f_1 f_2)[1-R(t)R_g(t)]}, \tag{5.45}$$

where f_1 and f_2 are the independent solutions of uniform equation (5.43) specified within $0 < h < H_s$, and

$$W = f_1\frac{df_2}{dh} - f_2\frac{df_1}{dh}. \tag{5.46}$$

The values R_g and R are defined as

$$R_g = -\frac{f_2'(0,t)+qf_2(0,t)}{f_1'(0,t)+qf_1(0,t)}, \tag{5.47}$$

$$R = -\frac{f_1'(H_s,t)-f_1(H_s,t)\dfrac{\varphi_1'(H_s,t)+R_i\cdot\varphi_2'(H_s,t)}{\varphi_1(H_s,t)+R_i\cdot\varphi_2(H_s,t)}}{f_2'(H_s,t)-f_2(H_s,t)\dfrac{\varphi_1'(H_s,t)+R_i\cdot\varphi_2'(H_s,t)}{\varphi_1(H_s,t)+R_i\cdot\varphi_2(H_s,t)}}, \tag{5.48}$$

$$R_i = \frac{\varphi_1'(H_i,t)-\varphi_1(H_i,t)\cdot\dfrac{\psi_1'(H_i,t)}{\psi_1(H_i,t)}}{\varphi_2'(H_i,t)-\varphi_2(H_i,t)\cdot\dfrac{\psi_1'(H_i,t)}{\psi_1(H_i,t)}}. \tag{5.49}$$

The functions φ_1 and φ_2 are independent solutions of the uniform equation (5.43) for $H_s < h < H_i$, and ψ_1 is the solution for $h > H_i$ which satisfies the condition

$$\frac{d}{dh}\arg\psi_1 > 0 \text{ with } h \to \infty. \tag{5.50}$$

The derivative sign f' in Eqs. (5.47)–(5.49) means the derivative over the h-variable. Apparently, R_g, R and R_i can be regarded as the reflection coefficients at $h = 0$, $h = H_s$ and $h = H_i$, respectively.

Now we can get to the hybrid representation of the received field, as manifested in the beginning of Section 5.2.

As a first step toward implementing this program we will express the factor $(1 - R_i R_g)^{-1}$ in terms of a partial geometric series,

$$(1 - R_i R_g)^{-1} = 1 + \sum_{n=1}^{N} (R_i R_g)^n + \frac{(R_i R_g)^{N+1}}{1-R_i R_g} \tag{5.51}$$

and substitute Eq. (5.51) into the integral (5.42):

$$W(\xi,h,h_0) = \sqrt{\frac{\xi}{\pi}}e^{-j\pi/4}\int_C dt e^{j\xi t}W^{-1}(f_1,f_2)\times \left\{ \begin{array}{l} F_1(h_0,t)[f_1(h,t)+Rf_2(h,t)]+ \\ \sum\limits_{n=1}^{N}(RR_g)^n F_1(h_0,t)[f_1(h,t)+Rf_2(h,t)]+ \\ \dfrac{(RR_g)^{N+1}F(h,h_0,t)}{1-RR_g} \end{array} \right\}. \tag{5.52}$$

Here

$$\begin{aligned} F_1(h_0,t) &= f_2(h_0,t) + R_g f_1(h_0,t), \\ F(h,h_0,t) &= -K(h,h_0,t)W(f_1,f_2)(1-RR_g) \end{aligned} \tag{5.53}$$

and N is an arbitrary integer whose choice is dictated by the length of the propagation path. It represents the maximum number of reflections from the upper "boundary" $h = H_i$ to be taken into account.

If there is no elevated M-inversion (cf. the dashed profile in Figure 5.6), the attenuation function $W(\xi, h, h_0)$ would take the form

$$W(\xi, h, h_0) = \sqrt{\frac{\xi}{\pi}} e^{-j\pi/4} \int_C dt e^{j\xi t} \frac{F_1(h_0, t)\left[f_1(h, t) + R_s f_2(h, t)\right]}{W(f_1, f_2)(1 - R_s R_g)}, \tag{5.54}$$

where R_s denotes the reflection coefficient at the "upper wall" $h = H_s$ of the surface duct, i.e.,

$$R_s = -\frac{f_1'(H_s, t) - f_1(H_s, t)\dfrac{\varphi_1'(H_s, t)}{\varphi_1(H_s, t)}}{f_2'(H_s, t) - f_2(H_s, t)\dfrac{\varphi_1'(H_s, t)}{\varphi_1(H_s, t)}}. \tag{5.55}$$

Of the two independent functions above H_s, i.e. φ_1 and φ_2, only φ_1 will remain because of the "new" condition

$$\frac{d}{dh} \arg\varphi > 0 \text{ with } h > H_s. \tag{5.56}$$

To isolate the set of eigenwaves of a solitary surface duct from the solution (5.52), we multiply and divide the first term of Eq. (5.52) by the "resonant denominator" of Eq. (5.54), i.e., $(1 - R_s R_g)$, and represent R as

$$R = R_s - R_i \frac{\Lambda_1 + R_s \Lambda_2}{1 + R_i \Lambda_2} = R_s - R_i \nu \tag{5.57}$$

with

$$\begin{aligned} \Lambda_1 &= \frac{f_1'(H_s, t)\cdot\varphi_2(H_s, t) - f_1(H_s, t)\cdot\varphi_2'(H_s, t)}{f_2'(H_s, t)\cdot\varphi_1(H_s, t) - f_2(H_s, t)\cdot\varphi_1'(H_s, t)}, \\ \Lambda_2 &= \frac{f_2'(H_s, t)\cdot\varphi_2(H_s, t) - f_2(H_s, t)\cdot\varphi_2'(H_s, t)}{f_2'(H_s, t)\cdot\varphi_1(H_s, t) - f_2(H_s, t)\cdot\varphi_1'(H_s, t)}. \end{aligned} \tag{5.58}$$

Now consider the function $g(h, t)$ defined as the solution of

$$\frac{d^2}{dh^2} g + [U_1(h) - t]g = 0 \tag{5.59}$$

with the conditions

$$\left(\frac{dg}{dh} + qg\right)_{h=0} = 0, \tag{5.60}$$

$$g(H_s, t) = C(t)\varphi_1(H_s, t) \tag{5.61}$$

where $U_1(h)$ corresponds to the dashed M-profile of Figure 5.6. We can obtain that

$$1-R_sR_g=\frac{\left(\dfrac{dg}{dh}+qg\right)_{h=0}}{g(0,t)\left[f_1'(0,t)+qf_1(0,t)\right]}\cdot \left[f_1(0,t)-\frac{f_1'(H_s,t)-f_1(H_s,t)\dfrac{\varphi_1'(H_s,t)}{\varphi_1(H_s,t)}}{f_2'(H_s,t)-f_2(H_s,t)\dfrac{\varphi_1'(H_s,t)}{\varphi_1(H_s,t)}}f_2(0,t)\right] \tag{5.62}$$

and hence Eq. (5.60) is equivalent to the characteristic equation

$$1-R_sR_g=0$$

specifying the modes of the surface duct. Further, we bring the integrand of Eq. (5.52) to the form

$$\frac{F_1(h_0,t)[f_1(h,t)+R_sf_2(h,t)]}{1-R_sR_g}-\frac{F_1(h_0,t)R_i\nu f_2(h,t)}{1-R_sR_g}-\frac{F(h,h_0,t)R_sR_g}{1-R_sR_g} \\ +\sum_{n=1}^{N}F(h,h_0,t)R_g^n\left[R_s^n+\sum_{p=1}^{n}\binom{n}{p}(-1)^pR_s^{n-p}R_i^p\nu^p\right]+\frac{(RR_g)^{N+1}F(h,h_0,t)}{1-RR_g} \tag{5.63}$$

with $\binom{n}{p}$ being binominal coefficients, and express the first integral as a sum of the residues at $t=\hat{t}_n$, or the zeros of the equation $1-R_sR_g=0$ (the bar serves to indicate their distinction from the roots of $1-RR_g=0$). Representing $(1-R_sR_g)^{-1}$ in the second and third term as another partial geometric series, generally speaking with a number of terms different from Eq. (5.51), we assemble similar terms and arrive at

$$W(\xi,h,h_0)=\sqrt{\frac{\xi}{\pi}}e^{-j\pi/4}(W_{01}+T_2+T_3+T_4) \tag{5.64}$$

where term W_{01} represents the sum of the normal modes of the surface duct

$$W_{01}=2\pi j\sum_{n=1}^{\infty}\exp\left(j\hat{t}_n\xi\right)\frac{d\hat{t}_n}{dq}\frac{g(h,\hat{t}_n)}{g(0,\hat{t}_n)}\frac{g(h_0,\hat{t}_n)}{g(0,\hat{t}_n)} \tag{5.65}$$

and

$$T_2=-\int_C dte^{j\xi t}W^{-1}(f_1,f_2)F_1(h_0,t)R_i\nu\cdot f_2(h,t), \tag{5.66}$$

$$T_3=\sum_{n=1}^{N}\int_C dt\frac{e^{j\xi t}}{W(f_1,f_2)}R_g^n\times \left[F(h,h_0,t)\sum_{p=1}^{n}\binom{n}{p}(-1)^pR_s^{n-p}R_i^p\nu^p-F_1(h_0,t)R_i\nu R_s^nf_2(h,t)\right], \tag{5.67}$$

$$T_4 = \int_C dt \frac{e^{j\xi t}}{W(f_1, f_2)} \left[\frac{F(h, h_0, t)}{1 - RR_g} \left(RR_g \right)^{N+1} - \frac{F_2(h, h_0, t)}{1 - R_s R_g} \left(R_s R_g \right)^{N+1} \right] \tag{5.68}$$

where

$$F_2(h, h_0, t) = F_1(h_0, t)[f_1(h, t) + R_s f_2(h, t)]. \tag{5.69}$$

The term T_2 can be interpreted as the field arriving at the observation point due to a single reflection from the elevated inversion while T_3 is a set of waves multiply reflected between the boundaries $h = H_i$, $h = H_s$, and $h = 0$. T_4 is a remainder term whose smallness will be secured by the proper choice of N for every specific separation from the source. It should be emphasized that in bringing Eq. (5.52) to the form of Eq. (5.64) we made no approximation, and hence Eq. (5.64) is an exact representation of the field in the structure analysed (to be more precise, it is exact in the same sense as the initial formula Eq. (5.42), which itself corresponds to a parabolic equation approximation). The set of normal modes associated with the surface duct, the entire infinite spectrum, has been separated from Eq. (5.42) solely by transforming the integrand (5.45). The fields given by T_2 and T_3 so far cannot be interpreted in the ray optical form, since rays appear only at the stage of asymptotic evaluation of the integrals.

As for the numerical truncation of the modal series, the question seems almost trivial, since in the absence of a reflecting upper wall the higher-order modes are evanescent and their respective eigenvalues have the ordering $\mathrm{Im}\hat{t}_{n+1} > \mathrm{Im}\hat{t}_n$.

As observed from Eqs. (5.66) and (5.67), the integrands T_2 and T_3 contain no singularities. Depending on the number of significant saddle points, each of these integrals splits into several terms representing waves which are radiated either up or down from the transmitting antenna. Further along the path, they are reflected from the boundaries $h = 0$, $h = H_s$ and $h = H_i$ to arrive at an observation point along some of the ray trajectories of Figure 5.7. For example, T_2 splits into two terms, each corresponding to a wave that is reflected once from the elevated layer. The first leaves the source in an upward direction, reaches a reflection point in the upper layer, and comes to the observation point along the ray 1, Figure 5.7. Another emerges downwards from the source downward, undergoes reflections from the earth's surface and the elevated layer, and arrives at the observation point along trajectory 2, Figure 5.7.

The different components of the angular spectrum of waves excited by the transmitter dipole are subject to a kind of spatial 'filtration' in the nonuniform troposphere. We can define the critical angle θ_0 as the angle at which the ray QO' emerging from the source turns at the height $h = H_s$, as shown in Figure 5.7. The rays with departure angles $\theta < \theta_0$ are trapped by the surface duct; those with $\theta > \theta_0$ leave it freely. The ray with $\theta = \theta_0 + \delta$, $\delta \to 0$ comes back to earth, upon reflection from the elevated layer, at the maximum range attainable to a single reflection from the elevated layer, we define this range as ξ_{1m}. Real rays arriving at greater distances than ξ_{1m} can do so only through a higher number of reflections from the "boundaries" $h = H_i$, $h = 0$ and $h = H_s$. Generally, each group of rays of a

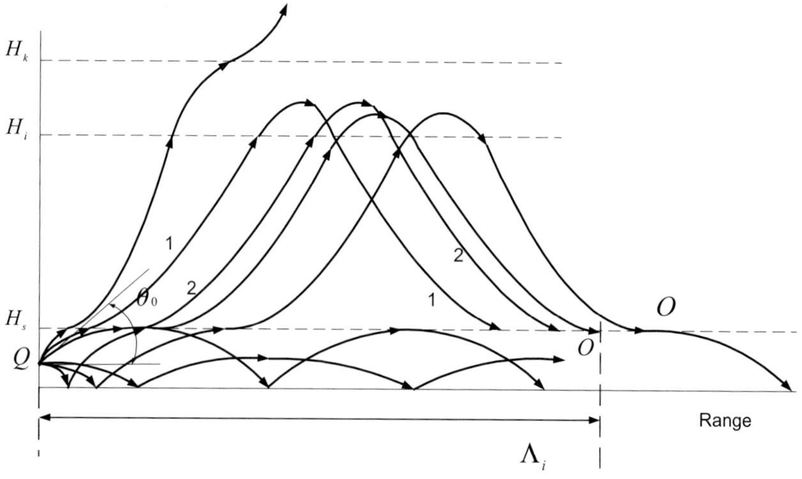

Figure 5.7 Schematic of the ray trajectories in the structure shown in Figure 5.6; Λ_i is the maximum range attainable through a single reflection from the upper layer.

given reflection multiplicity, l, has its own 'horizon', i.e. the limiting distance still corresponds to real stationary points in the integrand of T_2 (for $l = 1$) and/or T_3 ($l > 1$). It is in the vicinity of the lth horizon that the contribution of l times the reflected wave is the highest. Indeed, the grazing angle with respect to the elevated layer assumes the lowest possible value when the observation point is near the horizon. As a result, one can distinguish, along the path, characteristic zones of the "first hop" (i.e. near $\xi = \xi_{1m}$ where is the dominant contribution to $T_2 + T_3 + T_4$, for the N terms in T_3, only those with $l \le N$ 'work'). Moving beyond the lth horizon, an observer finds himself in an umbral zone where the field of the $l + 1$ hop has not yet formed while that of the lth hop can no longer propagate except by diffraction. The magnitude of this diffractional field is determined by the small contribution of complex rays corresponding to the complex stationary points of T_2 and/or T_3, but mainly by the value of the remainder integral T_4. For evaluating this latter it seems convenient to represent it as

$$T_4 = T_{41} + T_{42} \tag{5.70}$$

with

$$T_{41} = \int_{-\infty}^{\infty} dt \frac{e^{j\xi t} F_1(h_0,t)}{1-R_s R_g} \left[f_1(h,t) + f_2(h,t) \frac{R^{N+1} - R_s^{N+1}}{R^N - R_s^N} \right] R_g^{N+1} R(R^N - R_s^N), \tag{5.71}$$

$$T_{42} = \int_{-\infty}^{\infty} dt \frac{e^{j\xi t} F_1(h_0,t)(R_s R_g)^{N+1}}{(1-R_s R_g)(1-R R_g)} \left[f_2(h,t) + f_1(h,t) \frac{1}{R_s} \right] R(R - R_s). \tag{5.72}$$

As can be seen, T_{41} is the contribution of waves $N + 1$ times reflected from the upper layer, and hence it should be small near the horizon of $l \le N$-reflected waves.

The number of reflections, N, which are to be taken into account at a specific distance from the transmitter can be estimated as follows: Let us assume for simplicity that the reflection coefficient R_i depends on the grazing angle θ as

$$\begin{aligned} R_i(\theta) &= 1, \quad \theta \le \theta_0 \\ R_i(\theta) &= 0, \quad \theta > \theta_0 \end{aligned} \tag{5.73}$$

where θ is with respect to the lower edge of the layer, $r_i = a + Z_i$. Consider the field T_N due to reflections from the upper layer, for $h = h_0 = 0$ and $\xi \cong 2\sqrt{H_i}$ (i.e., near the horizon of singly reflected waves), i.e.,

$$T_N \sim \exp\left[j\xi t + j\frac{4}{3}N(H_i - t)^{3/2} - j\frac{4}{3}N(-t)^{3/2}\right]. \tag{5.74}$$

The saddle point of Eq. (5.74) is

$$t^{(N)} = -H_i\frac{\left(N^2-1\right)^2}{4N^2}. \tag{5.75}$$

Recalling the relation between the grazing angle and t, i.e.,

$$\cos\theta = \frac{ka+mt}{kr_i} \tag{5.76}$$

and the fact that real saddle points of real rays satisfy the inequality

$$t \le \min\,(h, h_0) \tag{5.77}$$

we can obtain from Eq. (5.76) an estimate at $t \to 0$ for the grazing angle θ of N times the reflected waves

$$\theta_{\min}^{(N)} \cong \left(\frac{2H_i}{a}\right)^{1/2}\frac{N^2+1}{2N}. \tag{5.78}$$

To have $R_i(\theta_{\min}) \neq 0$, it is necessary that

$$\left(\frac{2H_i}{a}\right)^{1/2}\frac{N^2+1}{2N} \le \theta_0\,. \tag{5.79}$$

The order of magnitude of θ_0 is 10^{-2}, and that of $(2H_i/a)^{1/2} \cong 5\times10^{-3}$ to 10^{-2}, whence the number of terms to be retained in Eq. (5.64) near the horizon $l = 1$ is $N = 1$–3.

We leave evaluation of the magnitude of term T_{42} in Eq. (5.72) until the next section, noting here that its contribution cannot be described in pure ray optical terms. In fact this term describes a peculiar physical effect of secondary excitation of the surface (evaporation) duct.

A better illustration of the relative importance of the different terms in Eq. (5.64) can be achieved through numerical analysis. Figure 5.8 represents the calculated attenuation function for $f = 10$ GHz in a two-channel structure of the refractivity. The surface M-inversion is characterised by $Z_s = 9$ m, $\Delta M = M(0) - M(Z_s) = 2$ M-units. The elevated M-inversion lies at the height $Z_i = 500$ m and has been approxi-

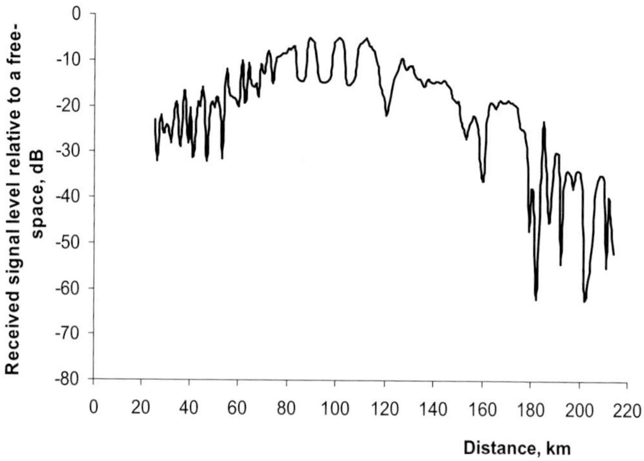

Figure 5.8 Calculated signal strength level for a two-channel system at frequency 10 GHz: $Z_s = 9$ m, $\Delta M = 2$ M-units, $Z_i = 500$ m and $\Delta M_i = 10$ M-units.

mated with a discontinuity $\Delta M_i = 10$ M-units of no thickness (i.e., $Z_k = Z_i$). An approximation like this obviously results in an overestimated value of the reflection coefficient $R_i(t)$. The heights of the transmitting and receiving antennas are $z_0 = 8.5$ m and $z = 5$ m, respectively, apparently both inside the evaporation duct. The normalised impedance q of the earth's surface is assumed to have a value corresponding to that of sea water: $\mathrm{Re}q = 6$, $\mathrm{Im}\ q = 94$. The calculated imaginary part of the first-mode eigenvalue is found to be $\mathrm{Im}\ t_1 = 0.4$. For a given combination of the antenna heights the optical horizon would be at ≈ 20 km. As observed from Figure 5.8, there is a region beyond that horizon (extending to about 60 or 70 km) where the field level is determined by the lower-order mode of the surface duct alone. Further along the path, the contribution of waves reflected from the elevated layer becomes substantial. In the zone of the first hop (i.e. 60 km $\le x \le$ 200 km) the maximum field strength becomes close to the free space level. The remainder term T_4 reaches a –40 dB level near distance 200 km with $N = 1$.

The mechanism of the radio wave propagation in the above two-channel system can be summarised as follows. At relatively smaller distances from the transmitter the dominant contribution to the received signal strength comes from evaporation duct propagation alone. Further along the path, the waves, departing with angles exceeding the critical angle θ_0 and reflected by the elevated M-inversion, contribute to the received field at an appropriate distance of the single hop. These waves interfere with each other as well as with the waves propagating through the evaporation duct producing the deep fades caused by large phase difference. For example, in Figure 5.8, the deepest fades are observed in the range 60–100 km where the wave propagating through the evaporation duct and superposition of the waves reflected from the elevated M-inversion are comparable in amplitude. Then, along the distance, the field reflected from the elevated M-inversion takes over and provides the major contribution to the received field strength.

5.2.1 Secondary Excitation of the Evaporation Duct by the Waves Reflected from an Elevated Refractive Layer

Referring to the discussion in the previous section and Figure 5.7, we state that the rays with departure angles from the source less than critical angle θ_0 are trapped in the surface duct, those with $\theta > \theta_0$ leave it freely. We also note that rays with $\theta > \theta_0$ reach the observation point at the distances $\xi < \xi_{1max}$. At distances greater than the horizon of first hop (ξ_{1max}, single reflection) there are no single reflected rays launched at the angles of departure $\theta > \theta_0$. Hence, between the first and second hop there is a range of distances where the receiving field is formed by another mechanism represented by term T_{42} in Eq. (5.72) and evaluated in this section. .

Consider the range of distances $\xi \geq \xi_{1max}$ concentrated on trapped waves/rays. The trapped waves with departure angles $\theta < \theta_0$ turn inside the surface duct, as shown in Figure 5.8. Since the potential barrier is finite the trapped waves can leak to the space above the surface duct as shown schematically in Figure 5.9, their respective phases are complex due to the leaking mechanism distributed along the distance. The field created by these waves in the space above the surface duct can be imagined as a cluster of rays distributed along the distance and sliding along the tangents to circles of radius $r = a + h_2$, where the height h_2 is given by the relation $2m^2\, 10^{-6}\, M(h_2) = t$ for $h_2 > Z_s$, where t is a dimensionless 'energy' corresponding to the angle θ.

In evaluating the maximum magnitude of T_{42} and therefore the impact from secondary excitation at $\xi \geq \xi_{1max}$ we shall set $N = 1$. Taking into account that the observation point lies within the surface layer, we will represent the integral (5.72) as a sum of residues at the zeros of $1 - R_s\, R_g = 0$.

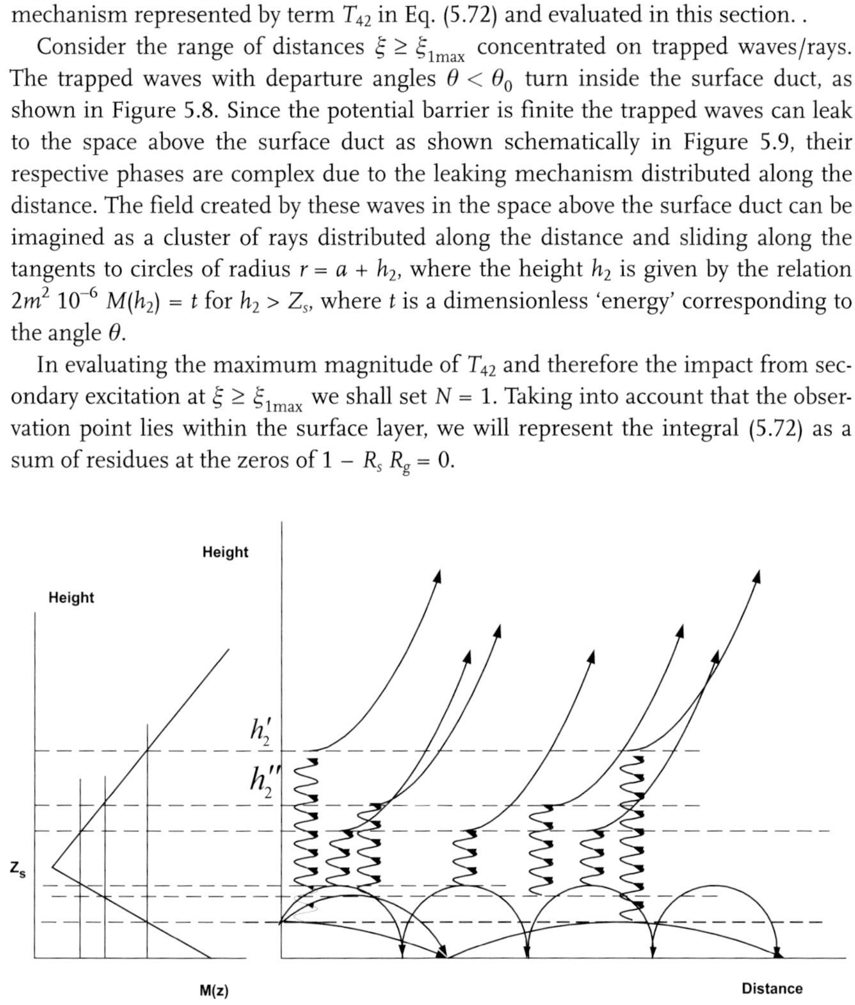

Figure 5.9 Schematic diagram of the penetration of "leaked" modes through the potential barrier in an evaporation duct.

Using Eq. (5.57) we can isolate the coefficient of reflection from the upper boundary of the surface duct R_s in R and present the denominator of Eq. (5.72) in the form

$$(1 - RR_g)(1 - R_s R_g) = (1 - R_s R_g)^2 \left[1 - \frac{R_i R_g \nu}{1 - R_s R_g}\right]. \tag{5.80}$$

The term ν may be interpreted as a coupling factor between the lower and upper ducts. For low attenuating modes ($\mathrm{Im} t_n << 1$) it can be estimated as

$$\nu(t_n) \approx \exp\left(-\frac{4}{3}\frac{\mu+1}{\mu} t_n^{3/2}\right) \tag{5.81}$$

where $\mu = |dU/dh|$ is a gradient of profile $U(h)$ for $0 < h < H_s$. The greater Re t_n the lower is the turning height h_1, the lesser the amount of leaking from the surface duct and, in accordance with a reciprocity theorem, such waves are harder to generate by a source distributed across the reflective layer but outside the surface duct.

To finish evaluation of T_{42}, we write a partial geometric series for $(1 - R_s R_g)^{-1}(1 - RR_g)^{-1}$ with the help of formula (5.80), namely,

$$(1 - R_s R_g)^{-1}(1 - RR_g)^{-1} = (1 - R_s R_g)^{-2} \times \left[1 + \sum_{p=1}^{P-1} \frac{(-R_i R_g \nu)^p}{(1 - RR_g)^p} + \frac{(-R_i R_g \nu)^P}{(1 - RR_g)(1 - R_s R_g)^{P-1}}\right] \tag{5.82}$$

and substitute it in Eq. (5.72).

Now we restrain the evaluation by the first term of the expansion (5.82) using the same arguments of angular filtration as in the previous section. Then we assume that $H_i - H_s >> 1$ and $a/H_i \cdot 10^{-6} \Delta M << 1$, where $\Delta M = M(z = 0) - M(Z_s)$ is the M-deficit of the surface duct. Therefore, we assume that the waves reflected from the elevated refractive layer can be regarded as a plain waves on their arrival at the level of the surface duct, and the horizons of departure of the leaked waves h_2 in Figure 5.9 are not to too different from H_s compared with H_i, i.e., $\max(h_2)/H_i << 1$. This allows us to use the same magnitude of the reflection coefficient $|R_i(t)| \approx |R_i(0)|$ for all rays from the cluster of leaking waves, expanding the amplitude and phase of the reflection coefficient into a series over powers of t. Taking into account that the observation point lies within the surface layer, we will represent the integral (5.72) as a sum of residues at the zeros of $(1 - R_s R_g)^2 = 0$

$$T_{42} \cong 2\pi j |R_i(0)| \exp\left\{j\frac{4}{3}\left[(H_i - H_s)^{3/2} + \delta(0)\right]\right\} \sum_{s=1}^{S} \exp(j\xi\hat{t}_s)(1+q)\nu(\hat{t}_s)\left(\frac{d\hat{t}_s}{dq}\right)^2 \times \left\{\begin{array}{l} \dfrac{g(h,\hat{t}_s)}{g(0,\hat{t}_s)}\dfrac{g(h_0,\hat{t}_s)}{g(0,\hat{t}_s)}\left[\dfrac{\nu'(\hat{t}_s)}{\nu(\hat{t}_s)} + j\xi + 3g^{-1}(0,\hat{t}_s)\dfrac{\partial g(0,\hat{t}_s)}{\partial \hat{t}_s} + \dfrac{d\hat{t}_s}{dq}\dfrac{\partial}{\partial \hat{t}_s}\left(\dfrac{d\hat{t}_s}{dq}\right)^{-1}\right] \\ + \dfrac{\partial}{\partial \hat{t}_s}\left[\dfrac{g(h,\hat{t}_s)}{g(0,\hat{t}_s)}\dfrac{g(h_0,\hat{t}_s)}{g(0,\hat{t}_s)}\right] \end{array}\right\}. \tag{5.83}$$

Here S is the number of modes trapped in the surface duct.

Equation (5.83) provides an estimate of the secondary excitation of the surface duct by a limited number of the modes of the surface duct, "trapped", from the point of view of geometrical optics but, in fact, "leaking" from the duct along the distance of propagation due to the finite thickness of the potential barrier in $U(h)$. These waves, then being reflected from the elevated M-inversion, reach the surface duct from above and "leak" back into the wave guide reproducing itself in the surface duct. Of course, the symmetry in the secondary excitation of the surface duct relies on the uniformity of the refractive index in the horizontal plane along the distance. Analysing Eq. (5.83) we may note that for the shorter wavelength λ of the electromagnetic radiation, the trapping condition of the surface duct becomes stronger and leaking of the trapped modes is reduced, as is the attenuation of the ducted field with distance. This leads to an even smaller contribution from the secondary excited field described by term T_{42}, Eq. (5.83). Increase in the wavelength λ leads, on the one hand, to a greater attenuation of the ducted field and stronger leaking of the trapped modes from the duct and, on the other hand, to better conditions for reflection from the elevated layer and stronger secondary excitation of the surface duct. The major contribution to the secondary guided field is from the modes with $|\mathrm{Re}\, t_n| << 1$. These modes have relatively high attenuation compared with deeply trapped modes, however they still provide a significant contribution at smaller distances from the source and, being reflected from the elevated layer, at distances $\xi \geq \xi_{1\max.}$ Evaluating this result numerically, we have found that the secondary guided field given by Eq. (5.83) may reach a maximum level of –50 dB relative to a free space condition with ideal reflection from the elevated layer, $|R_i| = 1$.

The rather exotic though observable situation is shown in Figure 5.10 where we have extremely strong elevated M-inversion and surface-based evaporation duct simultaneously. Here $Z_s = 18$ m, $\Delta M = M(0) - M(Z_s) = 6$ M-units, $Z_s = 500$ m, $Z_k =$

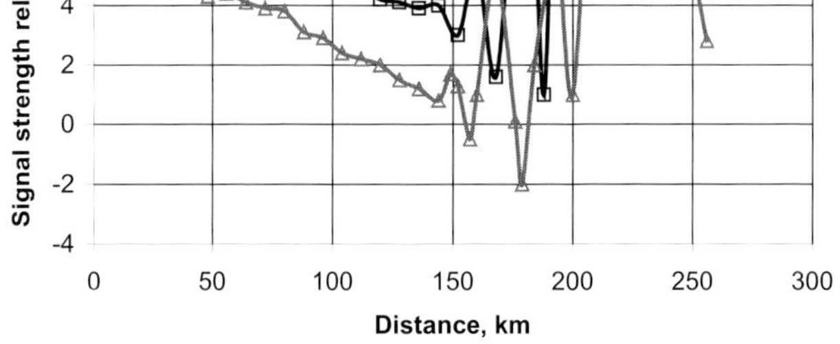

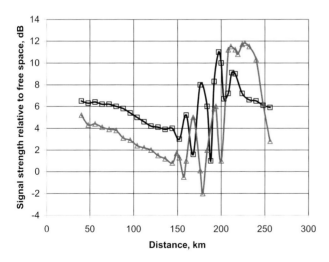

Figure 5.10 The range dependence of the attenuation factor in a two-channel system: □, both antennas located inside the evaporation duct; △, antennas above the evaporation duct.

600 m, $\Delta M_s = 60$ M-units. The calculation was performed for radio frequency $f = 3$ GHz and the following combination of antenna heights: $z_0 = 16$ m, $z = 10$ m and 50 m, respectivly.

Figure 5.10 shows that the field reflected from elevated M-inversion dominates at distances of the order of the reflection cycle length Λ_i despite there being lots of trapped modes of the elevated duct. At shorter distances the total field is a combination of the modes of the evaporation and elevated ducts with predominance of the evaporation duct. It also might be observed that at distances exceeding the first hop distance but still shorter than the second hop distance (in the above case with $x > 200$ km) the secondary excited field of the evaporation duct may provide a substantial contribution revealing a distinct change in the attenuation rate with distance and somewhat repeating the field at short range.

Figure 5.7 shows the ray trajectories for this situation. As observed, in the vicinity of the first hop the waves reflected from the elevated M-inversion provide a dominant contribution to the received field.

5.3 Comparison of Experiment with the Deterministic Theory of the Elevated Duct Propagation

Among numerous experiments we select Refs. [10–14] where the measured data have been compared with theoretical calculation of the propagation in the evaporation duct. In Refs. [10, 13], the authors studied the height structure of the received signal strength inside the elevated duct and good agreement between theory and measured data was observed for the location of the receiving antenna inside the duct. However, in the vicinity of the channel borders the measured data exceed the levels predicted from deterministic theory.

In Ref. [10], the measured data for frequencies 65, 170, 520 and 3300 MHz have been compared with the theory for the respective frequencies. The elevated M-inversion forms the single mode waveguide at the frequency 65 MHz. The characteristic values of the M-profile measured in Ref. [10] are as follows: $Z_i = 180$ m, $Z_k = 302$ m, $\Delta M_i = 40$ M-units; the waveguide channel is created in the interval of heights $0 < z < Z_k$, i.e., the duct belongs to the class of surface-based elevated ducts. Figure 5.11 from Ref. [10] shows very good agreement of the height dependence of the received signal strength with the theoretical calculations. It might be noted that in this frequency band the effects of scattering on the fluctuations of the refractivity are relatively small.

Figure 5.12 shows the measured data for 3.3 GHz in the same duct. As observed from the figure, at this frequency the elevated duct reveals the multimode structure of the received signal level. Inside the duct the measured signal strength levels exceed the level of the field due to the condition of "normal" refraction at 50–30 dB. The results of theoretical calculation (solid line) depicted in Figure 5.12 show good agreement with the measured data, fitting very well between the maximum and minimum signal levels observed in the experiment inside the duct. As seen from the same figure, the discrepancy between measured and predicted results becomes

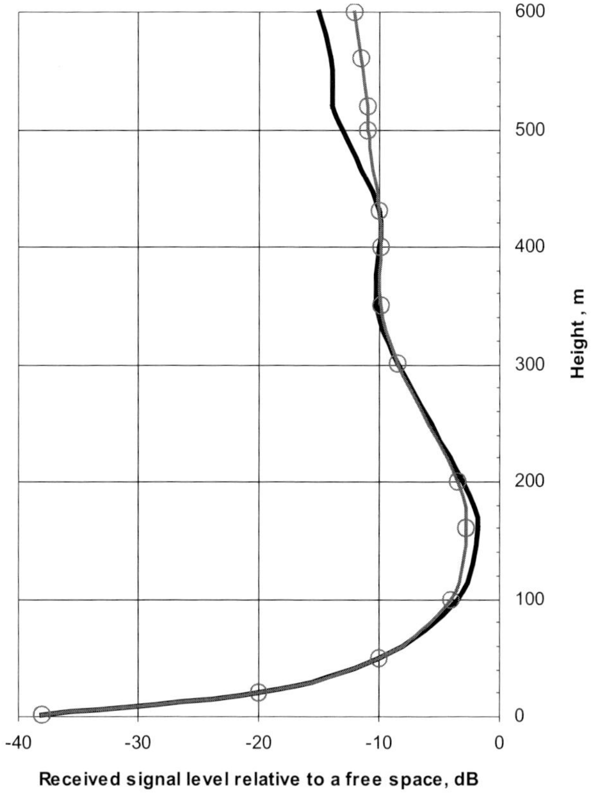

Figure 5.11 Height dependence of the received field level at a distance 111.2 km from the transmitter at a frequency of 65 MHz: solid line, measured data; ○, calculated using a split-step Fourier method.

significant when the receiving antenna are located close to boundaries inside the duct or outside the duct. The rapid decay of the calculated signal level at heights $z > 250$ m is caused by a destructive interference of many modes. At the same time the amplitude of each mode may decay at a lesser rate compared with the composite of all modes. As pointed out in Ref. [10], the phase relationship between modes may be distorted by non-uniformity of the refractivity structure in a horizontal plane.

The predicted path loss at frequency 3 GHz in Ref. [10] shows good agreement with the measured data for the transmitter and receiver located within the duct. Such good agreement with theory based on a rather crude model of stratified refractivity is likely due to the fact that slow variations in the refractive structure of the evaporation duct may result in "adiabatic" restructuring of the trapped modes and mutual transfer of energy between the resulting mode structure still largely formed by trapped modes. The result of this superposition in the received field strength would be close to one in the case of a horizontally uniform duct for the transmitter and receiver located far enough from the duct borders.

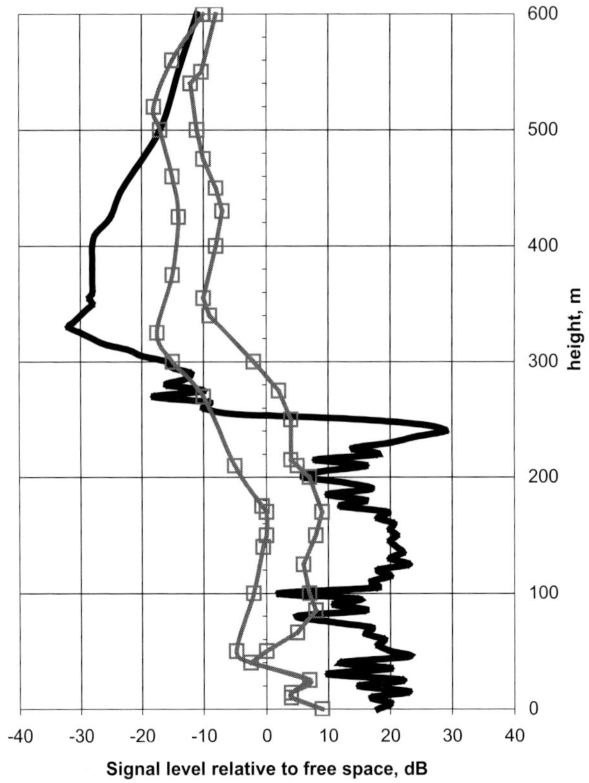

Figure 5.12 As Figure 5.11 but for frequency 3.3 GHz. Two solid curves with □ show minimum and maximum signal levels observed during the measurement interval.

Figure 5.13 from Ref. [10] shows the results of another experiment with an elevated duct detached from the ground surface. The parameters of the M-profile in this experiment were as follows: $Z_k = 800$ m, $z_{\min} = 600$ m, $\Delta M_i = 20$ M-units. The transmitting antenna was located at a height 20.7 m while receiving antenna was at a height 914 m above the sea surface, i.e. outside the elevated duct. With such placement of the receiving and transmitting antennas the trapped modes of the elevated duct cannot practically be generated by the radiating field of the transmitting antenna in such a configuration of the antennas in relation to the duct boundaries Z_k, $z_{\min}$ Figure 5.13 shows the result of the theoretical calculation of the received signal strength at frequency 3 GHz versus distance for a given placement of the antennas. As observed the calculated signal is most likely composed of the high order "leaked" modes with a rather great attenuation rate along the distance. At the same time, the measured signal level significantly exceeded that theoretically predicted from the deterministic theory of the stratified refractivity. The measured data also shown in Figure 5.13 do not attenuate with distance, behavior that is rather associated with trapped modes of the elevated duct.

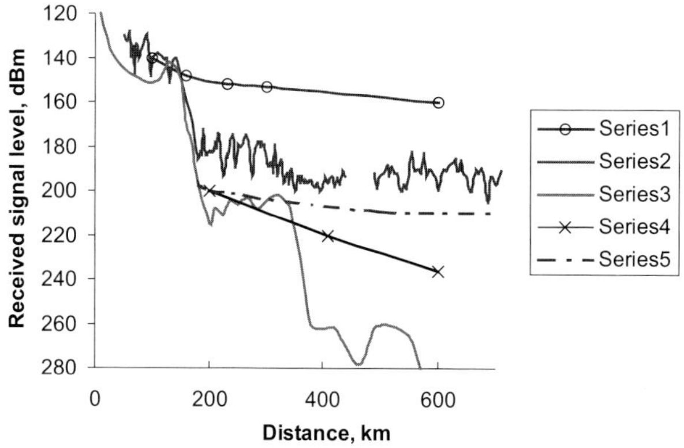

Figure 5.13 The received field strength vs. distance for elevated M-inversion at frequency 3.0877 MHz [10]: (1) signal strength in a free-space; (2) measured data; (3) theoretical results using averaged deterministic M-profile; (4) the level of the troposcattered signal; (5) results of the calculation using formula (5.90).

In order to evaluate another mechanism for long range radio wave propagation, the authors of Ref. [10] plotted in Figure 5.13 the level of the single-scattered field in accordance with the standard assumptions of the Booker–Gordon theory [15].

In a following study [13] the authors attempted to explain the observed signal levels in Figure 5.13 in the above experiment by the presence of an evaporation duct at the time of mthe easurements. Nonetheless, as discussed in Ref. [13], during the radio experiment there was no adequate measurement of the meteorological data which are required to restore the M-profile near the sea surface. The authors used instead some routine meteorological data available for that area and restored three possible M-profiles of the evaporation duct, one of which provides satisfactory agreement with the measured data in Figure 5.13.

It might be worthwhile to consider an alternative interpretation of the measured data in the above experiment, related to excitation of the trapped modes in the elevated duct due to scattering on the turbulent fluctuations of the refractive index inside the elevated duct. The respective theory and estimates are the subject of the next section.

5.4 Excitation of the Elevated Duct due to Scattering on the Fluctuations in the Refractive Index

Consider the piece-wise linear mode of the M-profile, introduced at the beginning of this chapter, Figure 5.1, and assume $Z_s = 0$, $\Delta M = 0$. In this section we investigate the mechanism of excitation of the elevated duct that can be formulated as follows:

The characteristic values of the thickness, $Z_k - Z_i$, of the elevated M-inversion and depth, ΔM_i, in most cases are themselves sufficient to ensure a waveguide mechanism for radio wave propogation in the range of 1–10 GHz of the frequency spectrum. As discussed in Chapter 1, the elevated duct location is defined in the interval of heights $Z_{\min} \le z \le Z_k$. The efficiency of the excitation of the guided waves nonetheless depends on the position of the transmitting antenna relative to the duct boundaries. If the antenna hight z_0 is outside the duct, say $z_0 < Z_{\min}$, it is "impossible", from the geometric optic point of view, to excite the guided waves. This means that the radiated field of the direct wave propagates through the elevated refractive layer experiencing slight refraction within, Figure 5.14. In this case the trapping of the waves inside the elevated duct is only possible due to scattering on the fluctuations of the refractive index, Figure 5.14.

The distances of interest are far beyond the horizon relative to the scattering volume, therefore we take into account only a single scattering into trapped modes of the elevated duct. For the same reason, the size of the scattering volume in the heights is limited by the interval of localisation of the trapped mode, i.e., $Z_{\min} \le z \le Z_k$. Let us assume that the width of the antenna pattern θ_a in azimuth is wide enough, i.e.,

$$\vartheta_a D >> L_{\|} \tag{5.84}$$

where D is the distance between the transmitter and the scattering volume, $L_{\|}$ is an external scale of the fluctuations in $\delta\varepsilon$ in the plane (x, y). In the current estimate of the scattering field we use an approximation for a free-space attenuation function W_0 of the direct wave. In this approximation for W_0 we do not take into account refraction (other than normal refraction) along the path between the transmitting antenna and a point within the scattering volume. In the problem of interest the impact of refraction is insignificant, since there is no trapping of the incident direct wave within the elevated refractive layer. We also will not take into account the effect of the boundary $z = 0$ on the scattered field. The presence of the interface $z = 0$

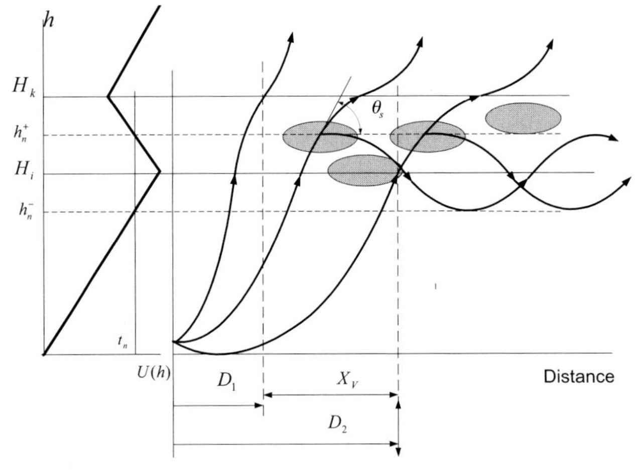

Figure 5.14 Schematic diagram of excitation of the elevated duct via a scattering mechanism.

leads to a lobed structure of the incident field. Such a complication, while easy to account for, is not worthwhile in the evaluation of the effect on the order of magnitude, which is the purpose of this investigation. Therefore, we take the artificial case of a grounded antenna $z =$ for further study in this section.

The Green function of the scattered field we define as the superposition (5.36) of the trapped modes of the elevated duct. The height-gain functions $\chi_n(z)$ are then defined by Eq. (5.38) within the scattering volume.

Another important assumption is that the major contribution to scattering comes from the inhomogeneities of $\delta\varepsilon$ in the vicinity of the upper boundary of the elevated refractive layer, i.e., with $Z_i \leq z \leq Z_k$. This assumption is founded on observations of the sharp increase in intensity of the fluctuation of $\delta\varepsilon$ in the vicinity of the upper boundary of the elevated layer [17], which is also associated with the upper boundary of the marine boundary layer. As a result, scattering into the trapped mode with number n is chiefly produced by the inhomogeneities located in the vicinity of the upper turning point $\overset{+}{z_n}$ of the wave associated with the nth mode, Figure 5.14:

$$\overset{+}{z_n} = \frac{m}{k\mu_3^3}\left(U_i - t_n - \mu_3^3 H_i\right). \tag{5.85}$$

In fact, the upper limit of integration η^+ over the height of the scattering volume for the nth mode is effectively limited by the height of the turning point $\overset{+}{z_n}$ and, from below, by the characteristic scale Λ_z of the oscillations in the height-gain function $\Lambda_z \sim m/k\mu_3$. Let us demand the following inequalities to be true

$$L_z << a/mt_n = \Lambda_x, \quad L_{\|} << X_V = D_2 - D_1 \tag{5.86}$$

where L_z is the characteristic external scale of the fluctuations in $\delta\varepsilon$. The parameter Λ_x in Eq. (5.89) is a characteristic scale of diffraction of the mode field due to the curvature of the earth, in the 10 GHz range $\Lambda_x \sim 10$ km; X_V is the length of the scattering volume, Figure 5.14, normally about several tens of km. Under condition

$$\theta_r << \theta_s \tag{5.87}$$

where $\theta_s = \lambda/L_z$ is the scattering angle in the vertical plane and $\theta_r = \sqrt{Z_i/2a}$ is the characteristic angle of refraction. The wave front of the incident wave has to be turned to an angle θ_r in order for that wave to be trapped in the elevated duct. The inequality (5.87) means that the angular width of the scattering exceeds the characteristic angle of refraction. In this case, the inequality

$$\frac{\Lambda_z}{2Z_i}\frac{\theta_r^2}{\theta_s^2} << 1 \tag{5.88}$$

holds and we can neglect the changes in the characteristic angle of refraction θ_r within the scattering volume. Let us estimate the common values involved in problem formulation: the height of the elevated layer Z_i is normally about $Z_i \sim 10^2 - 10^3$ m; the Fresnel zone size $\sqrt{\lambda D} \cong \lambda^{1/2}(2aZ_i)^{1/2}$ is about 100 m in the 10 GHz range; the external size $L_{\|}$ of the inhomogeneities $\delta\varepsilon$ in the horizontal plane is about $L_{\|} \approx 5 \times 10^2$–$10^3$ m [16] at heights of the order of Z_i; the vertical scale L_z is much smaller and is of the order of tens of meters because of the strong anisotropy present in vicinity the of the upper border of the atmospheric boundary layer [16]. We can also assume that

$$\Lambda_z >> L_z \tag{5.89}$$

which is most certainly satisfied for lower-order trapped modes, and then the intensity of the single-scattered field can be obtained in the form

$$J = \left\langle |W|^2 \right\rangle = 0.07k^{-2/3}\, C_\varepsilon^2 \alpha^{-5/3} (2m)^{11/3} \left(g_{23}/\xi_v \right) \times \sum_{n=1}^{N} \chi_n(z) e^{-\gamma_n \xi_v} \left(\xi_{2,n}^2 - \xi_{1,n}^2 \right) \left(h_n^+/\xi_n^+ \right)^{-8/3}. \tag{5.90}$$

Here $h_n^+ = kz_n^+/m$, $\xi_n^+ = (\xi_{1,n} + \xi_{2,n})/2$, $g_{23} = (\mu_2\mu_3)^6/(\mu_2^3 + \mu_3^3)^2$, C_ε is a structure constant of the fluctuations of $\delta\varepsilon$ in the vicinity of the upper border of the marine boundary layer $z \sim Z_i$. The parameters $\xi_{1,n}$, $\xi_{2,n}$ determine the non-dimensional borders of the effective scattering volume. To define them we use the geometry of the problem. Let us assume that the half-width of the antenna pattern in the vertical plane (E-plane) (x, z) is equal to θ_e. The ray which departs from the source at an angle θ_e can be related to the stationary value of the non-dimensional energy $t = m^2 E/k^2$ (see Section 2.2), equal to $t_1 = -m^2\theta_e^2$ (analog of the formula (2.52)). For the ray tangential to the earth's surface, as follows from Section 2.2, the stationary value is $t = t_2 = 0$.

The ray with stationary value t_1 reaches the upper $h_n^{(+)}$ and lower $h_n^{(1)}$ heights of the turning level for mode of number n and the respective distances

$$\xi_{1,n}^{(-)} = \frac{1}{\mu_2^3} \sqrt{t_n - t_1}, \tag{5.91}$$

$$\xi_{1,n}^{(+)} = \frac{1}{\mu_2^3} \sqrt{U_i - t_1} + \frac{1}{\mu_3^3} \left(\sqrt{U_i - t_1} - \sqrt{t_n - t_1} \right),$$

and the heights $h_n^{(+)}$, $h_n^{(-)}$ are given by

$$h_n^{(+)} = \frac{1}{\mu_3^3} \left(U_i - t_n + \mu_3^3 H_i \right), \tag{5.92}$$

$$h_n^{(-)} = \frac{1}{\mu_2^3} t_n .$$

The near boundary $\xi_{1,n}$ of the scattering volume in Eq. (5.90) can then be defined as $\xi_{1,n} = (\xi_{1,n}^{(-)} + \xi_{1,n}^{(+)})/2$. The far boundary $\xi_{2,n}$ can be defined in similar way using the Eq. (5.91) with t_1 replaced by $t_2 = 0$.

Curve (5) in Figure 5.13 shows the estimate of the scattered field in the elevated duct using Eq. (5.90) with the value of structure constant $C_\varepsilon^2 = 10^{-15}$ cm$^{-2/3}$. As observed from the figure, the strength of the estimated scattered field is 25 dB less than the level of the measured signal. This might be explained by underestimation of the intensity of the fluctuations in refractive index in the vicinity of tZ_k, which potentially can reach a value of two orders of magnitude higher than the value $C_\varepsilon^2 = 10^{-15}\ \mathrm{cm}^{-2/3}$ used here. Such values of C_ε^2 have been reported in Ref. [17]. Of course, any attempt to match the theory (5.90) with the measured data [10] would be speculative, since information about fluctuations in the refractive index was not reported in Ref.[10]. The main purpose of the above comparison is to give a somewhat different explanation of the mechanism of the propagation that can hardly be explained by the deterministic model of a stratified troposphere.

References

1 Wait, J.R. *Electromagnetic Waves in Stratified Media*, Oxford, Pergamon Press, Oxford, 1962, 372 pp.

2 Bremmer, H. *Terrestrial Radio Waves*, Elsevier, New York, 1949, 343 pp.

3 Dresp, M.R., Rather A.S. Tropospheric duct propagation beyond the horizon, *Proc. URSI Commission, Sect. F. Open Symp., La Baule, France*, 1977, Pt.1, pp. 31–36.

4 Baz, A.I., Zeldovich, Y.B., Perelomov, A.M. *Scattering, Reactions and Decays in. Nonrelativistic Quantum Mechanics*, Academic Press, New York, 1980.

5 Ott, R.H. Roots of the modal equation for em wave propagation in a tropospheric duct, *J. Math. Phys.*, 1980, 21 (5), 1256–1266.

6 Migliora, C.G. Felsen, L.B., Cho, S.H. High Frequency propagation in elevated tropospheric duct, *IEEE Trans. Antennas Propagation*, 1982, 30, 1107–1120.

7 Baumgartner, Jr., G.B., Hitney, H.V., Pappert R.A. Duct propagation modelling for the integrated refractive effects prediction system (IREPS)., *Proc. IEE*, 1983,130, Pt.F, 630–642.

8 Felsen, L.B. Hybrid ray-mode fields in inhomogeneous waveguides and ducts, *J. Acoust.* Soc. Am., 1981, 69 (2), 352–361.

9 Marcus, S.V. A model to calculate em fields in tropospheric duct environments at the frequencies through SHF, *Radio Sci.*, 1982, 17, 895–901.

10 Pappert, R.A., Goodhart, C.L. Case study of beyond-the-horizon propagation in tropospheric duct environments, *Radio Sci.*, 1977, 12 (1), 75–81.

11 Kukushkin, A.V., Sinitsin, V.G. Rays and modes in non-uniform troposphere, *Radio Sci.*, 1983, 18 (4), 573–581.

12 Guinard, N.W., Ransone, J., Randall, D. et al. Propagation through an elevated duct. Tradewind 3, *IEEE Trans. Antennas Propagation*, 1964, 12 (4), 479–453.

13 Hitney, H.V., Pappert, R.A., Hattan, C.P. Evaporation duct influences on beyond-the-horizon high altitude signals, *Radio Sci.*, 1978, 13 (4), 669–675.

14 Chang, H.T. The effect of tropospheric layer structures on long-range VHF radio propagation, *IEEE Trans. Antennas Propagation*, 1971, 19 (6), 751–756.

15 Booker, H., Gordon, W. A theory of radio scattering in the troposphere, 1950, *Proc. IRE*, 1950, 38 (4), 401–412.

16 Gavrilov, A.S., Ponomareva, S.M. *Turbulence Structure in the Ground Level Layer of the Atmosphere*, Collected Data, Meteorology Series, 1984, No.1, Research Institute for Meteorological Information, Obninsk, (in Russian).

17 Deardorff, J.W, Willis, G.E. Further results from a laboratory model of the convective boundary layer, *Bound. Layer Met.*, 1985, 35, 205–236.

6
Scattering Mechanism of Over-horizon UHF Propagation

As discussed in Chapter 1, the analytical study of wave propagation through a turbulent troposphere is complicated due to the high dynamic range of the turbulent fluctuations of the refractive index as well as the presence of the boundary surface.

In Chapter 4 we studied the impact of the random component of the dielectric permittivity of the troposphere on UHF propagation in an evaporation duct. It was shown that wave scattering on random inhomogeneities of the refractive index leads to a non-coherent redistribution of the energy between the waveguide modes, the loss of coherence results in additional attenuation of the trapped modes. Using the language of quantum mechanics, we have found the perturbations to the eigenvalues of the discrete spectrum of the localised states. The results obtained in Chapter 4 are valid at distances from the source, on the one hand, long enough to filter all leaked modes and, on the other hand, not too long so that the scattering in the upper layers of the troposphere can be neglected in comparison with the field carried by the trapped modes.

In the absence of tropospheric ducts it is sometimes convenient to solve the transport equations for the coherence function in the continuum spectrum. In Ref. [1] such a solution has been obtained for multiple wave scattering in a random medium (uniform on average) over the flat boundary surface. The results of Ref. [1] demonstrated that anisotropic inhomogeneities of the refractive index may have a significant impact on the signal level. The multiple scattering in the case of strong anisotropy was studied in Ref. [2], where the following was shown:

Compared with the case of isotropic fluctuations of $\delta\varepsilon(\vec{r})$, when the attenuation of the coherent component of the received field is defined by the square of the phase fluctuations, multiple scattering on anisotropic fluctuations with a factor of anisotropy $\alpha = L_{\perp}/L_{\parallel} << 1$ results in lesser attenuation of the coherent component by a factor of $(kL_{\perp}\alpha)$ with the same value of $L_{\parallel}$ and the scale of the frequency correlation increases $(kL_{\perp}\alpha)^{-2}$ times. Similar results were obtained in Ref. [2], where the authors derived integro-differential equations for the field moments which account for diffraction on the random inhomogeneties and variations in the trength of scattering with small scattering angles. The important result of Ref. [2] is that the Markov approximation is not applicable with extremely large parameters of anisotropy. In such cases the *two-scale model* [3] may provide a sufficient tool for analysis.

Radio Wave. Alexander Kukushkin

ISBN: 3-527-40458-9

In this chapter we suggest a technique for calculating the coherent signal component, with allowance for both diffraction by the earth and wave scattering on random inhomogeneities of the refractive index. Theoretically, the coherent component is defined as the complex amplitude averaged over a statistical ensemble of realisation. If the ergodicity and locally frozen hypotheses hold, the ensemble averaging is equivalent to such over an infinitely long interval. For practical purposes, however, the component coherent over a finite interval of time is of interest. According to data provided in Refs. [4–7], in about 60% of experiments performed on trans-horizon links up to 300 km in length, the received amplitude distribution is different from Rayleigh law. This difference is ascribed to the effect of the coherent component. Therefore, it seems necessary to analyse all the factors influencing the coherent component, its range and wavelength dependences, etc.

The approach can be formulated in the following way. For the electromagnetic field component, coherent over time T, we derive, using the Markov approximation, a parabolic-type equation allowing for the additional decay of the coherent signal due to scattering on small-scale fluctuations of the refractive index. Further, we analyse the possibility of replacing in the equation the average (over time T) dielectric constant, which is a "slowly varying" random function of all three variables, with a random function depending on a single coordinate, i.e. height over the interface. Through this procedure, the problem of wave propagation through a three-dimensional random medium is reduced to that for a random stratified medium. The field component averaged over time T can be represented in this approach as a normal wave (i.e. modal) series with random propagation constants and height gain functions for each mode. These are determined with the aid of the perturbation technique formulated in Chapter 3, in which the unperturbed refractive index profile is that averaged over the ensemble of the realizations, and the random stratification due to large-scale anisotropic inhomogeneities is considered as a perturbation. The approach permits one to obtain closed-form expressions for statistical moments of any order and analyse the correlation between the signal levels and the turbulent troposphere.

It seems noteworthy that at decimetre wavelength the attenuation rate of the coherent component ($T \leq 1$ min) is not normally very high, hence at links about 200 km long the coherent intensity practically coincides with the whole decimetre signal. The basic observation is that the approach suggested in this chapter is better suited to wave propagation effects at frequencies below 1 GHz, where the effect of the ducting is also week.

6.1
Basic Equations

Consider a vertically polarized field whose attenuation function (2.16) is governed by the parabolic equation

$$2j\frac{\partial W}{\partial x} + \Delta_{\perp} W + k^2[\varepsilon_M(x, y, z) - 1]W = 0 \tag{6.1}$$

where $\varepsilon_M(x, y, z) = \varepsilon(x, y, z) + 2z/a$.

As is known, scattering on the fluctuation in the dielectric permittivity brings about additionnal attenuation of the coherent component of the field which can be described by the factor $e^{-\gamma x}$. In the Markov process approximation, the decrement of attenuation γ is given by [8]

$$\gamma = \frac{\pi k^2}{4} \int d^2\vec{\kappa}\Phi_\varepsilon(0, \vec{\kappa}_\perp), \text{ with } \vec{\kappa} = \{\kappa_x, \kappa_y\}.$$

Substitution of a Kolmogorov turbulence spectrum in the above equation would result in a divergence of the integral at small $\vec{\kappa}$. The reason for this is that the ensemble average usually implied in theoretical calculations allows for arbitrarily large phase distortions due to very large inhomogeneities or over an infinitely long time interval. In practice, however, the coherence over a finite time interval and the dependence of the coherent amplitude on the time of average is of interest.

Let us average Eq. (6.1) over a finite time T, denoting the average values as $\langle \ldots \rangle_T$, and make use of the locally "frozen", stationary turbulence hypothesis [9]. This hypothesis implies that variations of the random field $\varepsilon(\vec{r}, t)$ with time result solely from the motion of the turbulent flow at a velocity $\vec{v}$ which can be a random value itself. Introducing $\delta\varepsilon_T = \varepsilon_M(\vec{r}) - \langle \varepsilon_M(\vec{r}) \rangle_T$ and transforming Eq. (6.1) to the integral equation form we obtain

$$\begin{aligned}\langle W(\vec{r})\rangle_T = {} & W(0,y,z)\left\langle \exp\left(j\frac{k}{2}\int_0^x dx'\delta\varepsilon_T\left(x',y,z\right)\right)\right\rangle_T \\ & -\frac{1}{2jk}\left\langle \int_0^x dx' \exp\left(j\frac{k}{2}\int_{x'}^x dx''\delta\varepsilon_T\left(x'',y,z\right)\right)\cdot\left[\Delta_\perp + k^2\left[\langle\varepsilon_M(\vec{r})\rangle_T - 1\right]\right]W(\vec{r})\right\rangle_T .\end{aligned} \tag{6.2}$$

Instead of averaging the functionals depending on $\delta\varepsilon_T$ over time T, one can perform averaging over the ensemble of ε as $\delta\varepsilon$ does not contain greater time-scales than T. On the other hand, the term $\langle \varepsilon_M(\vec{r}) \rangle_T$ may be considered invariant over times $t \leq T$. Assuming statistical independence of the fluctuations in $\delta\varepsilon_T$ and $\vec{v}$, we can perform averaging over $\vec{v}$ independently of $\delta\varepsilon_T$. Suppose we wish to first perform the averaging over $\delta\varepsilon_T$. Introduce definition $W_T \equiv \langle W(\vec{r}) \rangle_T$ and then, as has been shown in Ref. [8], the term W_T will obey the equation

$$2j\frac{\partial W_T}{\partial x} + \Delta_\perp W_T + \left[k^2\left(\langle\varepsilon_M(x,y,z)\rangle_T - 1\right) - 2jk\gamma_T\right]W_T = 0 \tag{6.3}$$

provided that the wave propagation can be regarded as a Markovian process, i.e. the conditions below are met

$$L_{\|} << kL_z^2,\ \sigma_\varepsilon^2 k^2 L_{\|}^2 << 1. \tag{6.4}$$

Here, $L_{\|} = \sqrt{L_x^2 + L_y^2}$, L_z are, respectively, the horizontal and the vertical scales of the inhomogeneities and σ_ε^2 is the root mean square magnitude of fluctuations in ε.

The attenuation rate γ_T averaged over the fluctuations in Eq. (6.3) is given by the relation

$$\gamma \approx \frac{\pi k^2}{4} \int_{|\kappa_y|>\kappa_{yT}} d\kappa_y \int_{\kappa_z^2>\kappa_{zT}^2} d\kappa_z \Phi_\varepsilon(0, \kappa_y, \kappa_z) \tag{6.5}$$

and

$$\kappa_{yT} = (v_y T)^{-1}, \quad \kappa_{zT} = \left(\int_0^T (T-\tau) B_{vz}(\tau) d\tau \right)^{-1/2}; \tag{6.6}$$

v_y can be estimated as a mean value of the horizontal transfer velocity $v_\| = \sqrt{v_x^2 + v_y^2}$, while the vertical component of the velocity $\vec{v}$ can be regarded as solely random with the correlation function B_{vz}. We can assume, for further estimates, that $\kappa_{zT} \approx (\sigma_{vz} T)^{-1}$. The conditions (6.4) were obtained in Ref. [8] for a spatially uniform field of $\delta\varepsilon$. The inequalities (6.4) impose some limitations to an averaging time T, since under "frozen" turbulence conditions $L_\| = v_\| T, \quad L_z = \sigma_{vz} T$, therefore instead of Eq. (6.4) we obtain

$$\frac{k^2 \sigma_{vz}^2 T}{v_\|} << 1, \quad k^2 C_\varepsilon^2 v_\|^{8/3} T^{8/3} << 1 \tag{6.7}$$

and

$$\sigma_\varepsilon^2 = 1/2\, C_\varepsilon^2 L_\|^{2/3}$$

for "locally uniform" turbulence.

The function $\langle \varepsilon_M(\vec{r}) \rangle_T$ involved in Eq. (6.3) is a much slower function of x and y than $\varepsilon_M(\vec{r})$, at least with sufficiently long averaging time. It should be noted that vertical variations of the averaged dielectric permittivity are much more rapid than the horizontal-plane variations [9,10].

Establish at which times of averaging the x and y dependence of $\langle \varepsilon_M(\vec{r}) \rangle_T$ can be totally neglected in Eq. (6.3). In other words, for which T the term $\langle \varepsilon_M(\vec{r}) \rangle_T$ can be replaced by a vertical profile $\varepsilon_{Ti}(z)$ related to some fixed (generally, arbitrary) values $x = x_i$ and $y = y_i$, i.e.

$$\varepsilon_{Ti}(z) = \langle \varepsilon(x_i, y_i, z) \rangle_T. \tag{6.8}$$

The set of functions $\varepsilon_{Ti}(z)$ can be regarded as an ensemble of realisations of some random function $\varepsilon_T(z)$. Averaging over that ensemble will be denoted hereinafter by a horizontal bar above the character, e.g. $\bar{\varepsilon}_T(z)$. The substitution of $\varepsilon_T(z)$ instead of $\langle \varepsilon_M(\vec{r}) \rangle_T$ corresponds to the introduction of a vertically stratified random medium whose "instantaneous" profile $\varepsilon_T(z)$ is the same along the entire propagation path. The slow variations of $\langle \varepsilon_M(\vec{r}) \rangle_T$ are equivalent to altered realisations of $\varepsilon_T(z)$. Obviously, the transition of the vertically stratified medium can only be justified if the amount of fluctuations (over time T) in the coherent component that are due to the difference between $\langle \varepsilon_M(\vec{r}) \rangle_T$ and $\varepsilon_T(z)$ is small. Introducing

$$\varepsilon_1(x, y, z) = \langle \varepsilon_M(x, y, z) \rangle_T - \varepsilon_T(z) \tag{6.9}$$

we can present $W_T(x, y, z)$ as

$$W_T(\vec{r}) = W_T^0(\vec{r}) \exp(\Psi(\vec{r})) \tag{6.10}$$

where $\Psi = V + jS$ is a complex phase and $W_T^0(\vec{r})$ satisfies the equation

$$2j\frac{\partial W_T^0}{\partial x} + \Delta_\perp W_T^0 + \left[k^2(\varepsilon_T(z) - 1) - 2jk\gamma_T\right] W_T^0 = 0. \tag{6.11}$$

Let $W_T^0(\vec{r}, \vec{r}_0)$ denote the propagation factor of the field from a point source at $\vec{r}_0 = \{0, 0, z_0\}$. Within the first-order approximation of Rytov's method, the complex phase of Eq. (6.10) is [8]:

$$\Psi = -\frac{k^2}{W_T^0(\vec{r}, \vec{r}_0)} \int_0^x dx' \int d^2\vec{\rho}' G\left(\vec{r}, \vec{r}'\right) \varepsilon_1\left(\vec{r}\right) W_T^0\left(\vec{r}', \vec{r}_0\right), \tag{6.12}$$

where

$$\vec{\rho}' = \left\{x', y'\right\}, \vec{r}' = \left\{x', \vec{\rho}'\right\}, G\left(\vec{r}, \vec{r}'\right) = -\frac{1}{4\pi(x - x')} W_T^0\left(\vec{r}, \vec{r}'\right). \tag{6.13}$$

We will consider the case when the reception point $\vec{r}'$ is shadowed from the transmitter by the terrestrial sphere. The complex phase fluctuations represented by Eq. (6.12) depend essentially on the vertical profile of $\varepsilon_1(\vec{r})$. If $\varepsilon_1(\vec{r})$ contains spectral components characterised by small vertical scales $l_z << m/k$, then the mechanism of trans-horizon propagation is a resonance scattering in the higher tropospheric layers, i.e. the troposcatter described by the Booker and Gordon theory [11]. In this case the fluctuations in the log-amplitude V are very high.

To reduce the standard of fluctuations in the log-amplitude V determined by Eq. (6.10) it is necessary to suppress the multi-scale random variations of $\varepsilon_1(\vec{r})$ by choosing a sufficiently long averaging time T. The minimum vertical scale sizes of the $\varepsilon_1(\vec{r})$ inhomogeneities which are related to the average time as $l_z = \sigma_{vz} T$ should meet the condition $l_z >> m/k$. The resultant condition on the averaging time is

$$T >> \frac{m}{k\sigma_{vz}}. \tag{6.14}$$

In contrast to the "standard" troposcatter mechanism [11], the scattering of waves satisfying the above conditions occurs in the region of the troposphere which is in the shadow zone with respect to both receiver and transmitter. Hence, the propagation factor $W_T^0(\vec{r}, \vec{r}_0)$ and the Green function $G\left(\vec{r}, \vec{r}'\right)$ are most conveniently represented as modal series, Chapters 2 and 3. Rough estimates for $W_T^0(\vec{r}, \vec{r}_0)$ and $G\left(\vec{r}, \vec{r}'\right)$ values can be obtained by retaining just the first terms of these expansions, i.e.

$$W_T^0\left(\vec{r},\vec{r}'\right) = -2\exp\left(j\frac{\pi}{4}\right)\frac{m}{k}\sqrt{\frac{\pi m\left(x-x'\right)}{a}}\frac{\chi(z)\chi(z')}{N} \times$$
$$\exp\left\{jq\frac{x-x'}{2k}+j\frac{k}{2}\frac{\left(y-y'\right)^2}{x-x'}-\gamma_T\left(x-x'\right)\right\}. \tag{6.15}$$

Estimating the integral in Eq. (6.12) we will first assume $\varepsilon_T(z)$ to correspond to the standard tropospheric refraction, i.e. $\varepsilon_T(z) = 1 + 2\,z/a$. Explicit expressions for the values involved in Eq. (6.15) are

$$q = \left(k^2/m^2\right)\tau_1 e^{j\pi/3},\ \tau_1 = 2.338,\ N = -4\,m/k\,\sqrt{\tau_1}e^{j\pi/3}\ \text{and}$$
$$\chi(z) = w_1\left(\tau_1 e^{j\pi/3} - kz/m\right),$$

where w_1 is the Airy function defined in the Appendix. Performing the integration in Eq. (6.15) and averaging over the ensemble of $\varepsilon_1(\vec{r})$ on the assumption of statistically independent fluctuations thereof, we can arrive at the following results.

With

$$T >> \frac{\sqrt{\lambda x}}{v_{\|}} \tag{6.16}$$

the mean-square log-amplitude fluctuations are

$$\left\langle V^2\right\rangle = 4\cdot 10^{-4}\pi^{3/2}C_{\varepsilon_1}^2 L_{\|}^{-7/3}x^3 \tag{6.17}$$

where C_{ε_1}, is the structure constant of ε_1. The structure function has been assumed to obey the "two-thirds" law, i.e. the spectrum of ε_1 takes the form (1.34). Thus, provided the following inequalities hold,

a) $4\cdot 10^{-4}\pi^{3/2}C_{\varepsilon_1}^2 v_{\|}^{-7/3}T^{-7/3}x^3 << 1,$

b) $k^2 C_\varepsilon^2 v_{\|}^{8/3}T^{8/3} << 1,$ (6.18)

c) $T\max\left\{\frac{m}{kv_z},\frac{\sqrt{\lambda x}}{v_{\|}},\frac{v_{\|}}{k\sigma_{vz}^2}\right\},$

the term $\langle\varepsilon_M(x,y,z)\rangle_T$ in Eq. (6.3) can be replaced by $\varepsilon_T(z)$ without introducing considerable error in $W_T(\vec{r})$. Substituting $C_{\varepsilon_1}^2 \approx C_\varepsilon^2 = 10^{-14}$ cm$^{-2/3}$, $\lambda = 10$ cm, $T = 30$ s, $v_{\|} = 10$ m s^{-1}, $\sigma_{vz} = 1$ m s^{-1}, we see that the principal inequality (6.18a) can hold at $x \le 10^3$ km.

6.2
Perturbation Theory: Calculation of Field Moments

Thus, calculation of the coherent field amplitude averaged over time T has been reduced to solving Eq. (6.11). If $\varepsilon_T(z)$ is a regular function, Eq. (6.11) can be solved by known methods, with the solution presentable in the form (6.14). Actually, the problem consists in determining the propagation constants q and the height-gain functions $\chi(z)$. Depending on the specific form of $\varepsilon_T(z)$, this can be done either analytically or numerically. However, this is a rare occasion. Generally, $\varepsilon_T(z)$ is a random function whose only known parameters are the vertical stratification scale L_z and the root mean square fluctuation strength $\left\langle \varepsilon_T(z)^2 \right\rangle^{1/2}$. We further assume that averaging over the statistical ensemble is equivalent to averaging over time, $\left\langle \varepsilon_T^2 \right\rangle = \overline{\varepsilon_T^2}$ i.e. the validity of the ergodicity theorem. An analytical solution to Eq. (6.11) cannot be written, we should aim instead at evaluating the mean level of $W_T(\vec{r})$ and the root mean square fluctuations $\sigma_W = \sqrt{\overline{W_T^2}}$ thereof from knowledge of the $\varepsilon_T(z)$ statistics. Averaging over the statistical ensemble we can write

$$\varepsilon_T(z) = \overline{\varepsilon_T(z)} + \Delta\varepsilon(z), \quad \overline{\Delta\varepsilon} = 0 \tag{6.19}$$

assuming specifically

$$\overline{\varepsilon_T(z)} = \varepsilon_0(z) = 1 + 2\frac{z}{a}. \tag{6.20}$$

To determine the propagation constants and height gain functions, we will make use of the perturbation theory described in Section 3.1. Representing q and χ as

$$q = q_0 + \delta q, \tag{6.21}$$

$$\chi(z) = \chi_0(z) \exp\left(\int_0^z dz' \zeta\left(z'\right) \right), \tag{6.22}$$

where q_0 and χ_0 are governed by the unperturbed Eq. (6.11) with $\varepsilon_T(z) = \varepsilon_0(z)$. We can obtain the random correction term δq and $\zeta(z)$ in the form

$$\delta q = -\frac{k^2}{N^2} \int_0^\infty \Delta\varepsilon(z) \chi_0^2(z) dz, \tag{6.23}$$

$$\zeta(z) = \frac{1}{\chi_0^2(z)} \int_0^z \left[\delta q - k^2 \Delta\varepsilon\left(z'\right) \right] \chi_0^2\left(z'\right) dz'. \tag{6.24}$$

Now we single out of $W_T(x, y, z; z_0)$ the value W_0, i.e. the solution to Eq. (6.11) with $\varepsilon_T(z) = \varepsilon_0(z)$, to obtain

$$W_T(x, y, z; z_0) = W_0(x, y, z; z_0) \cdot \exp\left\{ j\frac{\delta q}{2k} x + \int_0^z \zeta\left(z'\right) dz' + \int_0^{z_0} \zeta\left(z'\right) dz' + \right\}. \tag{6.25}$$

The factor $W_T(x, y, z; z_0)$ is a random function of a "slow" time $t > T$. Within each realization of the signal, the factor W_T is the coherent component of the total signal received during the "short" intervals $t \leq T$. During the "long" time-interval $t >> T$, the factor W_T undergoes relatively slow random variations owing to changes of realisation of ε_T. The part of W_T fluctuating at the change of realisations $\Delta\varepsilon(z)$ is given by the exponent of Eq. (6.25).

We can specify the intensity J_0 of the coherent component $W_T(x, y, z; z_0)$ given $\varepsilon_0(z)$ in the form (6.20):

$$J_0 = |W_0|^2 = \frac{\pi m x}{4a\tau_1}\left|w_1\left(\tau_1 e^{j\pi/3} - \frac{kz}{m}\right)\right|^2 \left|w_1\left(\tau_1 e^{j\pi/3} - \frac{kz_0}{m}\right)\right|^2 \times \exp\left(-\frac{mx}{a}\tau_1\sqrt{3} - 2\gamma_T x\right) \tag{6.26}$$

where $2\gamma_T$ is the extra attenuation rate due to the energy transfer to the incoherent part of the signal. Generally, the attenuation rate γ_T given by Eq. (6.5) undergoes random variations as the medium realisation changes, because of the non-uniformity in the small scale fluctuations of $\varepsilon_M(\vec{r})$. In the following study, we shall restrict ourself to the case where the $\varepsilon_M(\vec{r})$ fluctuations of scale sizes $l_\| < v_\| T$ and $l_z < v_z T$ are statistically uniform, i.e. $\gamma_T = const.$

As can be observed from Eqs. (6.23) to (6.25), the random value W_T should be distributed log-normally, in view of the central limiting theorem. Note that similar distributions are actually observed in the experiment. For instance, according to data reported in Ref. [5], the integral distribution of the amplitude measured over 1 to 5 min intervals reveals log-normal statistics.

To simplify further the derivations, we will assume the receiver and transmitter heights to be equal, i.e. $z = z_0$, and consider the intensity J_T of the coherent (over time T) signal component averaged over the statistical ensemble, viz.

$$\overline{J_T} = \left\langle |W_T|^2 \right\rangle = J_0 \exp\left\{ \begin{array}{l} -\dfrac{x^2}{8k^2}\left\langle (\delta q - \delta q^*)^2 \right\rangle + \\ j\dfrac{x}{2k}\left\langle (\delta q - \delta q^*) \int\limits_0^z \left[\zeta(z') + \zeta^*(z')\right] dz' \right\rangle + \\ 2\int\limits_0^z dz' \int\limits_0^z dz'' \left\langle \left[\zeta(z') + \zeta^*(z')\right]\left[\zeta(z'') + \zeta^*(z'')\right] \right\rangle \end{array} \right\}. \tag{6.27}$$

Further calculations are straightforward but cumbersome. As a specific example, consider one of the correlators involved in Eq. (6.27):

$$\begin{aligned} \int\limits_0^z dz' \int\limits_0^z dz'' \left\langle \zeta(z')\zeta(z'') \right\rangle &= \left\langle (\delta q)^2 \right\rangle \left[\int\limits_0^z \frac{dz'}{\chi_0^2(z')} \int\limits_0^{z'} dz'' \chi_0^2(z'') \right] - \\ & 2\frac{k^4}{N} \int\limits_0^z \frac{dz'}{\chi_0^2(z')} \int\limits_0^{z'} dz_1 \chi_0^2(z_1) \int\limits_0^z \frac{dz''}{\chi_0^2(z'')} \int\limits_0^{z''} dz_2 \chi_0^2(z_2) \int\limits_0^\infty dz_3 \chi_0^2(z_3) \langle \Delta\varepsilon(z_2)\Delta\varepsilon(z_3) \rangle + \\ & k^4 \int\limits_0^z \frac{dz'}{\chi_0^2(z')} \int\limits_0^z \frac{dz''}{\chi_0^2(z'')} \int\limits_0^{z'} dz_1 \int\limits_0^{z''} dz_2 \chi_0^2(z_1)\chi_0^2(z_2) \langle \Delta\varepsilon(z_1)\Delta\varepsilon(z_2) \rangle. \end{aligned} \tag{6.28}$$

Other terms in the exponent of Eq. (6.27) can be expressed in a similar way.

After lengthy but straightforward calculations, somewhat simplified with $z << m/k\sqrt{\tau_1}$, we can arrive at

$$\overline{J_T} = J_0 \exp\left\{ \begin{array}{l} -\frac{x^2}{8k^2}\left\langle (\delta q - \delta q^*)^2 \right\rangle + j\frac{xz^2}{6k}\left[\left\langle (\delta q)^2 \right\rangle - \left\langle (\delta q^*)^2 \right\rangle\right] + \\ j\frac{kxz^4}{90m^2}\left[\tau_1 e^{-j\frac{\pi}{3}}\left\langle (\delta q^*)^2 \right\rangle - \tau_1 e^{j\frac{\pi}{3}}\left\langle (\delta q)^2 \right\rangle\right] - \\ \frac{kx\,\tau_1\sqrt{3}}{m^2\;90} z^4 \left\langle |\delta q|^2 \right\rangle + \frac{z^4}{18}\left\langle (\delta q + \delta q^*)^2 \right\rangle \end{array} \right\}. \tag{6.29}$$

The corresponding terms δq can be expressed via the Fourier transform of $\Delta\varepsilon(z)$, viz.

$$\tilde{\varepsilon}(\kappa) = \frac{1}{2\pi}\int_{-\infty}^{\infty} dz e^{-j\kappa z}\Delta\varepsilon(z), \tag{6.30}$$

$$\delta q = -k^2 \int_{-\infty}^{\infty} d\kappa \cdot \tilde{\varepsilon}(\kappa) V(\kappa) \tag{6.31}$$

where the function $V(\kappa)$ is given by

$$V(\kappa) = \frac{1}{N}\int_0^{\infty} dz \cdot \chi_0^2(z) e^{j\kappa z}. \tag{6.32}$$

It has the meaning of the scattering coefficient from the first mode back to itself again, due to inhomogeneities $\Delta\varepsilon(z)$ of the scale-size $l = 2\pi/\kappa$.

According to the inequality (6.14), the major contribution to the random component of the vertically stratified $\Delta\varepsilon(z)$ is given by sufficiently large inhomogeneities with $\kappa = \kappa_z << k/m$. For this case $V(\kappa)$ can be represented by an asymptotic expansion

$$V(\kappa) = 1 + j\frac{m\kappa}{k}\tau_1 e^{j\pi/3} + O\left(\left(\frac{m\kappa}{k}\right)^2\right). \tag{6.33}$$

Defining the spectral density of the fluctuations as

$$\Phi_1(\kappa) = \sqrt{2\pi}\sigma_{\Delta\varepsilon}^2 L_z \exp\left(-\frac{\kappa^2 L_z^2}{2}\right), \tag{6.34}$$

with $L_z = \sigma_{vz} T$, we shall substitute Eq. (6.33) into Eq. (6.29) with an account of the spectrum in the form (6.34). Then the average intensity $\overline{J_T}$ becomes

$$\overline{J_T} = J_0 \exp\left\{ \frac{3\pi x^2}{4L_z^2} m^2 \sigma_{\Delta\varepsilon}^2 \tau_1^2 - x\frac{k^5 z^4 \sigma_{\Delta\varepsilon}^2 \tau_1 \sqrt{3}}{180 m^2} - \frac{4}{3}\pi\sigma_{\Delta\varepsilon}^2 (kz)^4 \right\}. \tag{6.35}$$

The range of validity of Eq. (6.35) is dictated, according to the perturbation method employed, by the demand that the second-order correction to $W_T(x, z; z_0)$ be small.

Analyzing Eqs. (6.27) and (6.29) we find that the corrections to the propagation constants take the major role in the second-order corrections overall, and these can be evaluated as

$$\delta q^{(2)} \approx \frac{m^2}{k^2}(\delta q)^2. \tag{6.36}$$

Noting that $\delta q \sim k^2 \sigma_{\Delta\varepsilon}$ we will demand that the terms containing the range (distance) squared in the exponent are small, whence

$$\sigma^3_{\Delta\varepsilon} k^2 x^2 m^2 << 1. \tag{6.37}$$

Substituting $\sigma^2_{\Delta\varepsilon} \approx 10^{-13}$, we find that the requirements of Eq. (6.37) can be met for $x \leq 150$ km at $\lambda = 10$ cm ($f = 3$ GHz) and $x \leq 700$ km at $\lambda = 30$ cm ($f = 1$ GHz).

Shown in Figure 6.1 are the range dependences of the average intensity $\overline{J_T}$ as calculated from Eq. (6.35) for $f = 3$ GHz and $f = 1$ GHz. The parameter values assumed for the calculation are $z = z_0 = 10$ m, $\sigma^2_{\Delta\varepsilon} \approx 10^{-13}$, $L_s = 20$ m and $a = 8500$ km. Range dependences J_0 are also shown for comparison. As can be seen from Figure 6.1 and from Eq. (6.35), the presence of random gradients $d\varepsilon_T/dz \sim \sigma_{\Delta\varepsilon}/L_z$ in the refractive index results in a sharp increase in the signal strength. This can be regarded as a kind of "trapping" or localization of the radiated field near the earth's surface, however, in this case it is of a random nature.

Making use of Eqs. (6.25), (6.27) and (6.35), we can evaluate the square of the standard of the intensity fluctuations, viz.

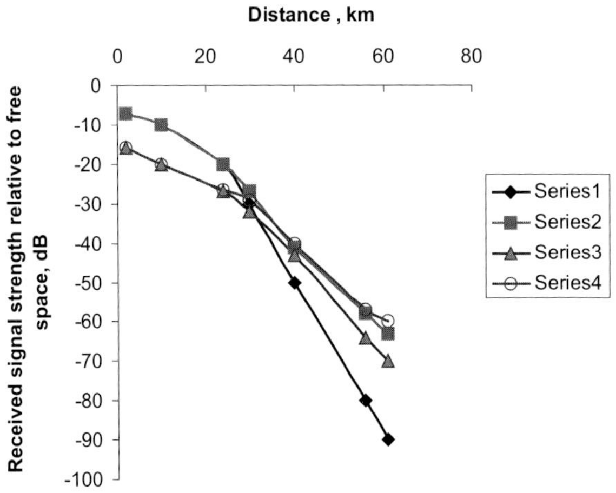

Figure 6.1 Range dependence of the average intensity of the received field, $10\log(\overline{J_T})$: (1) Received field strength at frequency 3 GHz in the absence of fluctuations in the refractivity ($\sigma^2_{\Delta\varepsilon}=0$); (2) received field strength at frequency 3 GHz, $10\log(\overline{J_T})$ in the presence of fluctuations in the refractivity $\sigma^2_{\Delta\varepsilon}\approx 10^{-13}$; (3) received field strength at frequency 1 GHz in the absence of fluctuations in the refractivity($\sigma^2_{\Delta\varepsilon}=0$); (4) received field strength at frequency 1 GHz, $10\log(\overline{J_T})$ in the presence of fluctuations in the refractivity $\sigma^2_{\Delta\varepsilon}\approx 10^{-13}$.

$$\sigma_J^2 = \overline{J_T^2} - \overline{J_T}^2 = \overline{J_T}^2 \left\{ \exp\left[\frac{3\pi x^2}{2L_z^2} m^2 \sigma_{\Delta\varepsilon}^2 \tau_1^2 - x \frac{k^5 z^4 \sigma_{\Delta\varepsilon}^2 \tau_1 \sqrt{3}}{90 m^2} - \frac{2}{9} \pi \sigma_{\Delta\varepsilon}^2 (kz)^4 \right] - 1 \right\}. \tag{6.38}$$

Figure 6.2 provides range dependences of the fluctuation standard $\sigma_J^2(x)$and the variation index $\beta_J^2(x) = \sigma_J^2(x)/\overline{J_T}^2(x)$calculated for the same parameter values. The latter magnitude β_J^2 characterizes the relative fluctuations in the intensity (i.e. the depth of "slow" fading). As observed from Eq. (6.38) and from Figure 6.2, the mean square magnitude of the intensity fluctuations is proportional to the average intensity and grows with the range as the first and major term in the exponent (6.38).

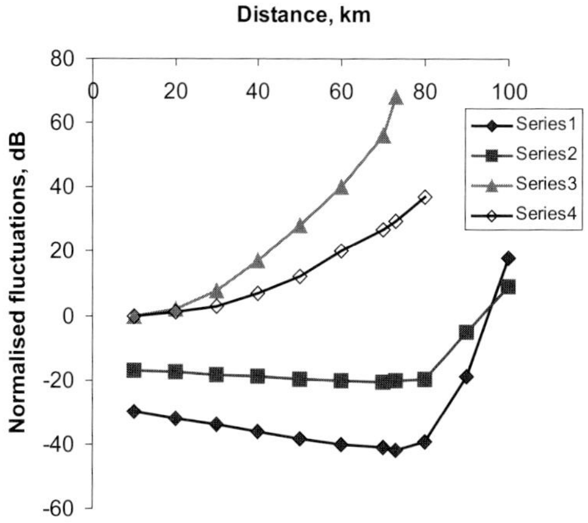

Figure 6.2 Range dependences of the fluctuation standard and the variation index in the presence of fluctuations in the refractivity with $\sigma_{\Delta\varepsilon}^2 \approx 10^{-13}$: (1) Fluctuation standard $\sigma_J^2(x)$ at frequency 3 GHz; (2) fluctuation standard $\sigma_J^2(x)$ at frequency 1 GHz; (3) variation index $\beta_J^2(x)$at frequency 3 GHz; (4) variation index $\beta_J^2(x)$at frequency 1 GHz.

We believe that Eqs. (6.35) and (6.38) can provide an explanation for the increase in the fading depth which is observed experimentally with increase in the mean levels of the signal [5]. According to the field measurements discussed in Refs. [4, 5], the fading depth increases with the path length up to ~ 200 km, such behavior is also in agreement with the theoretical result provided by Eq. (6.38). The above theory, developed in this section, introduces a two-scale model of fluctuations in the refractive index: small-scale fluctuations treated as a Kolmogorov turbulence and large-scale fluctuations $\Delta\varepsilon$ which are treated as a random stratification $\Delta\varepsilon(z)$.

Within the limits of this theory, the signal strength should demonstrate strong correlation with the fluctuation standard $\sigma_{\Delta\varepsilon}^2$ of the dielectric permittivity near the earth's/ocean's surface. As $\sigma_{\Delta\varepsilon}^2$ increases, the field intensity should grow, even with small gradients of the average profile of the refractive index, i.e. without ducting or enhanced refraction. This increase in the signal strength is due to multiple scattering of the lowest propagation mode on the "layered" inhomogeneities whose vertical scale sizes are commensurate with $\Lambda_z = m/k$, the vertical scale size of the mode oscillations, as shown in Figure 6.3. A one-dimensional analogy (along the z-coordinate) to this phenomenon is the stochastic parametric resonance considered in Ref. [12].

To analyse cases of ducted radio wave propagation in an atmospheric duct over the sea's surface, the common approach is to attempt to compare the measured data with theoretical predictions for $\varepsilon_M(z)$profiles recovered from meteorological measurements. To a certain degree, controlled by the validity of the hydrodynamic theory of an evaporation duct [13], such profiles correspond to the ensemble averaged dependences $\overline{\varepsilon_T(z)}$, with the random stratification $\Delta\varepsilon(z)$ apparently disregarded. Yet, as can be seen from the above theory, the random component of $\varepsilon(z)$ can play the dominant part in cases where the average profile does not reveal a strong near-surface inversion, like an evaporation duct or for frequencies significantly below 10 GHz when an evaporation duct, even if present, is normally insufficient for the ducting mechanism at these frequencies.

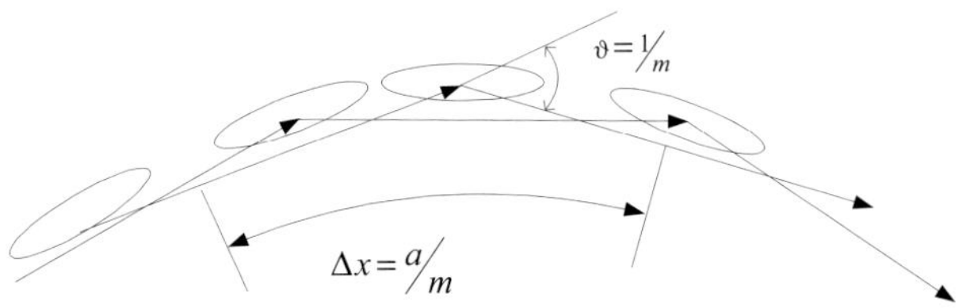

Figure 6.3 Schematic representation of multiple scattering of a diffracted field.

6.3 Scattering of a Diffracted Field on the Turbulent Fluctuations in the Refractive Index

Long-range tropospheric propagation due to re-emission of the energy of electromagnetic waves by inhomogeneities of the refractive index has been known since the 1940s [4]. The simple mechanism of the single scattering was first developed by Booker and Gordon [12] and then updated taking into account the Kolmogorov theory of the turbulent spectrum of fluctuations in the refractive index [9]. The theory of single scattering was proposed to explain the phenomenon of the long-range propagation in the absence of super-refractive anomalies in the refractive index profiles. It takes into account scattering by inhomogeneities located in the region formed by the intersection of the directional diagrams of the receiving and transmitting antennas, as shown in Figure 6.4.

According to Ref. [12], the mean intensity J_s of the scattered field I_s normalised on the intensity in a free space I_{FS} is expressed as follows:

$$J_s = \frac{I_0}{I_{FS}} = 16\,\sigma_0 \frac{V}{R^2} \tag{6.39}$$

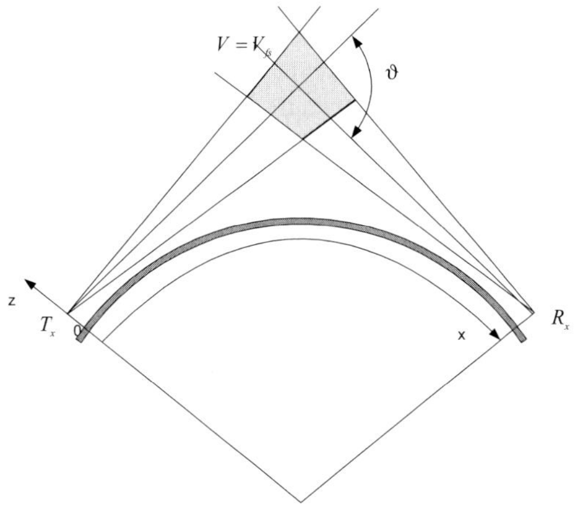

Figure 6.4 Geometry of single scattering in the troposphere.

where V is the effective scattering volume and R is the distance between the receiver and transmitter, and σ_0 is the effective scattering cross-section.

The effective scattering cross-section σ_0 from a unit volume to a unit solid angle is given by

$$\sigma_0 = \frac{\pi k^2}{2} \Phi_\varepsilon(\vec{q}), \tag{6.40}$$

where $\vec{q}$ is the scattering vector and $\Phi_\varepsilon(\vec{q})$ is the spectral density of the fluctuations in the dielectric permittivity $\delta\varepsilon(\vec{r})$. The scattering vector $\vec{q}$ has the components $q_x = q_y = 0$ and $q_z = 2k \sin(\vartheta/2)$, where x and y are coordinates along the surface of the earth and z is along the normal to it, k is the wavenumber and ϑ is the scattering angle, Figure 6.4.

For the inertial interval of the turbulence spectrum s_0 is determined by the relation

$$\sigma_0 = 0.052 k^{1/3} C_\varepsilon^2 \left(2 \sin \frac{\vartheta}{2}\right)^{-11/3}, \tag{6.41}$$

where C_ε is a structure constant, Section 2.

The above Booker–Gordon theory provides rather good estimates of the signal strength of the scattered field as well as describing the range dependence of the receiving signal strength. However, there are several factors which, in the majority of observations, are not in agreement with the theory of single scattering, for example, the dependence on wavelength, elevation angle, and the cumulative distribution of the scattered field [5, 7].

In this section we attempt to estimate the intensity of the signal over the horizon due to scattering of the waves in the volume located in the geometric shadow relative to both the transmitter and receiver, as shown in Figure 6.5.

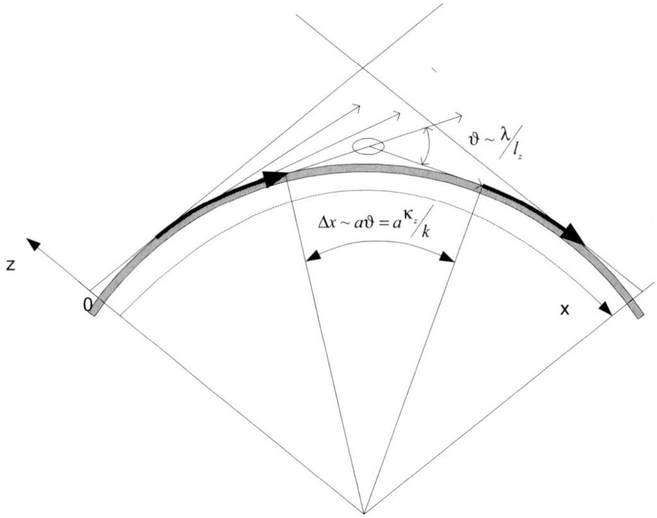

Figure 6.5 Geometry of the single scattering of the diffracted field.

Consider the case of normal refraction, in which the modified dielectric constant of the troposphere $\varepsilon_M(x, y, z) = \varepsilon_0(z) + \delta\varepsilon(x, y, z)$, $\varepsilon_0 = 1 + 2z/a$. Here $\delta\varepsilon(x, y, z)$ is a random component of the dielectric permittivity, $\langle\delta\varepsilon\rangle = 0$.

We shall define the mean intensity J_s of the scattered field normalised at an intensity in a free space via the attenuation factor W:

$$J_s = \frac{\langle I_s\rangle}{I_{FS}} = \left\langle |W|^2 \right\rangle. \tag{6.42}$$

The attenuation factor can be represented in the form $W = W_0 + W_1$, where W_0 is the solution of the equation

$$2jk\frac{\partial W_0}{\partial x} + \frac{\partial^2 W_0}{\partial y^2} + \frac{\partial^2 W_0}{\partial z^2} + k^2\left[\varepsilon_0(z) - 1\right] W_0 = 0. \tag{6.43}$$

and W_1 is defined in the Born approximation by the expression

$$\begin{aligned} W_1 = -k^2 \int_{x_0}^{x} dx' \int_{-\infty}^{\infty} dy' \int_{0}^{\infty} dz' G\left(x - x', y, z; y', z'\right) \times \\ \delta\varepsilon\left(x', y', z'\right) W_0\left(x' - x_0, y', z', y_0, z_0\right) \end{aligned} \tag{6.44}$$

where $\vec{r}$ and $\vec{r}_0$ are the vector-coordinates of the receiver and transmitter respectively, $\vec{r} = \{x, y, z\}$, $\vec{r}_0 = \{x_0, y_0, z_0\}$. Later we will set $x_0 = 0$ and $y = y_0 = 0$.

We also restrict the integration volume for discussion of the scattering of a diffracted field in the shadow region by the inequalities:

$$z' < \min\left\{\frac{\left(x' - \Delta x_0\right)^2}{2a}, \frac{\left(x - x' - \Delta x\right)^2}{2a}\right\}, \tag{6.45}$$

$$\Delta x_0 < x' < x - \Delta x.$$

Here $\Delta x_0 = \sqrt{2az_0}$, $\Delta x = \sqrt{2az}$. We note that the Booker–Gordon theory takes account of scattering only in the radiated region bounded by the inequality

$$z' > \left\{ \frac{(x' - \Delta x_0)^2}{2a}, \frac{(x - x' - \Delta x)^2}{2a} \right\}. \tag{6.46}$$

We can define the Green function $G(\vec{r}, \vec{r}')$ in terms of the attenuation factor,

$$G\left(\vec{r}, \vec{r}'\right) = \frac{1}{4\pi\left(x - \vec{x}\right)} W_0\left(\vec{r}, \vec{r}'\right), \tag{6.47}$$

which is represented in shadow region in the series of normal waves, Section 2,

$$W_0\left(\vec{r}, \vec{r}'\right) = -2e^{j\pi/4}\left[\frac{\pi m\left(x - x'\right)}{a}\right]^{1/2} \frac{m}{k} \exp\left\{j\frac{k\left(y - y'\right)^2}{2\left(x - x'\right)}\right\} \times$$
$$\sum_{n=1}^{\infty} \exp\left\{jq_n \frac{x - x'}{2k}\right\} \frac{\chi_n(z)\chi_n(z_0)}{N_n} \tag{6.48}$$

where q_n is a complex propagation constant

$$q_n = \frac{k^2}{m^2}\tau_n e^{j\pi/3}, \quad \tau_n = \left[\frac{3}{2}\pi\left(n - \frac{1}{4}\right)\right]^{2/3}, \quad N_n = -4\frac{m}{k}\sqrt{\tau_n}e^{j\pi/3}. \tag{6.49}$$

The height-gain functions $\chi_n(z)$ satisfy the equation

$$\frac{d^2\chi_n}{dz^2} + \left[k^2(\varepsilon_0(z) - 1) - q_n\right]\chi_n = 0 \tag{6.50}$$

and boundary conditions

$$\chi_n(z = 0) = 0, \quad \frac{d}{dz}\arg\chi_n(z) > 0 \quad \text{with} \quad z \to \infty. \tag{6.51}$$

For the case of normal refraction $\chi_n(z)$ can be represented in terms of the Airy function, $\chi_n(z) = w_1(t_n - \mu z)$, where $\mu = k/m$ and $t_n = q_n/\mu^2$.

We shall substitute Eqs. (6.47) and (6.48) into Eq. (6.44) and take into account in the double summation over the modes only diagonal terms which do not oscillate with distance. Integrating in Eq. (6.44) over dy and assuming statistical uniformity of the fluctuations in the refractive index and their isotropy in the x-, y-planes, we can obtain the equation

$$\left\langle |W_1|^2 \right\rangle_n = 2\frac{(\pi k)^2 m}{ax} F(z, z_0) \exp\left(-\frac{mx}{a}\tau_n\sqrt{3}\right) \times$$
$$\int_0^{x_1/2} dx' \left[\left(x' - \Delta x_0\right)^2 + \left(x - \Delta x - x'\right)^2\right] \int_{-\infty}^{\infty} d\kappa_y \int_{-\infty}^{\infty} d\kappa_z \Phi_\varepsilon(0, \kappa_y, \kappa_z) \left|V_n(\kappa_z, x')\right| \tag{6.52}$$

for the intensity of the scattered field in the nth mode $\left\langle |W_1|^2 \right\rangle_n$, where $F(z, z_0) = \frac{m^2}{k^2|N_n|^2} |\chi_n(z)|^2 |\chi_n(z_0)|^2$ is a factor describing the dependence of the mean intensity of the scattered field on altitude and $x_1 = x - \Delta x - \Delta x_0$.

The function $V_n(\kappa_z, x')$ in Eq. (6.52) has the meaning of a coefficient of re-scattering from the nth mode into the nth mode by inhomogeneities with the scale $l_z = 2\pi/\kappa_z$ and is defined by

$$V_n(\kappa_z, x) = \frac{1}{N_n} \int_0^{x^2/2a} dz \chi_n^2(z) \exp(j\kappa_z z). \tag{6.53}$$

With $\kappa_z < \kappa/m\tau_n$ and $x >> (a/m) \cdot \sqrt{\tau_n/2}$ the major contribution to integral (6.53) comes from the altitude's interval $0 < z < m\tau_n/2k$. The upper limit in Eq. (6.53) in this case can be replaced by infinity by means of introducing a smooth limiting function which compensates the growth of $\chi_n^2(z)$ at $z \to \infty$. Let us introduce such a function in the form of exponent $\exp(-\beta z)$. After transition to non-dimensional variables $\alpha = \beta\, m/k$, $\zeta = kz/m$ and $q = \kappa_z m/k$ we can deform the contour of integration over ζ in $V_n(\kappa_z, x)$ into a ray with $\arg \zeta = \pi/3$. With $\alpha \neq 0$ we can neglect the contribution from integration over the arc of infinite radius, and for $V_n(\kappa_z, x)$ we then obtain

$$V_n(\kappa_z, x) \approx V_n(\kappa_z) = \frac{1}{\sqrt{\tau_n}} \int_0^\infty d\zeta v^2(\zeta - \tau_n) e^{-q^* \zeta}, \tag{6.54}$$

where $q^* = (\alpha - jq)e^{j\pi/3}$, $v(\zeta - \tau_n)$ is the Airy function. We can further put parameter $\alpha = 0$ and expand the exponent into a series over powers of $q^*\zeta$

$$V_n(\kappa_z) = \frac{1}{\sqrt{\tau_n}} e^{-q^* \tau_n} \sum_{m=1}^{\infty} P_m \frac{(q^*)^m}{m!} (-1)^m, \tag{6.55}$$

where

$$P_m = \int_{-\tau_n}^{\infty} dx \cdot x^m v^2(x). \tag{6.56}$$

The recurrent formulas for integrals (6.56) are provided in the Appendix.

With $|q|\tau_n << 1$, we can retain only the first term in series (6.55). Taking into account that $v(-\tau_n) = 0$, we obtain the asymptotic expression

$$V_n(\kappa_z) \cong 1 + \frac{j}{3}\left(\frac{\kappa_z m \tau_n}{k}\right) \exp\left(j\frac{\pi}{3}\right) + O\left(\left(\frac{\kappa_z m \tau_n}{k}\right)^2\right)$$

which determines the contribution of non-resonant scattering by inhomogeneities having vertical scales l_z greater than the maximum characteristic scale $m\tau_n/\kappa$ of oscillations in $\chi_n^2(z)$.

When $k\tau_n^2/m << \kappa_z << 2kx/a$, the stationary point of the integrand in Eq. (6.53): $z_{st} = \frac{m}{k}\left(\frac{\tau_n}{2} + \left(\frac{\kappa_z m}{2k}\right)^2\right)$ makes the main contribution to $V_n(\kappa_z, x)$. In this case the resonant scattering of the nth mode occurs at the altitude z_{st} at which the scale of oscillations of $\chi_n^2(z)$ is equal to the vertical scale of the inhomogeneities, and we have for $V_n(\kappa_z, x)$ the expression

$$V_n(\kappa_z, x) \cong \frac{1}{2}\left(\frac{k}{\kappa_z m \tau_n}\right)^{1/2} e^{j\pi/12} H(z_s) \exp\left\{-j\frac{\kappa_z m t_n}{k} - j\frac{1}{12}\left(\frac{\kappa_z m}{k}\right)^3\right\}, \qquad (6.57)$$

$$H(z_s) = \int_{-\infty}^{z_s} \exp\left(j\xi^2\right) d\xi, \quad z_s = \frac{1}{4\sqrt{\kappa_z}}\left(\frac{m}{k}\right)^{3/2}\left[\left(\frac{2kx}{a}\right)^2 - \kappa_z^2\right]. \qquad (6.58)$$

When $\kappa_z > 2kx/a$, the upper limit of integration makes the main contribution to the integral $V_n(\kappa_z, x)$:

$$V_n(\kappa_z, x) \cong \frac{1}{2}\left(\frac{a}{2mkx\tau_n}\right)^{1/2} e^{j\pi/12} \exp\left\{-j\frac{mxt_n}{a} - j\frac{2}{3}\left(\frac{mx}{a}\right)^3\right\}. \qquad (6.59)$$

This equation corresponds to non-resonant scattering by inhomogeneities with scales $l_z < \pi a/kx$.

Let us substitute into Eq. (6.52) the asymptotes (5.57) and specify the spectrum of fluctuations in the refractive index $\Phi_\varepsilon(\vec{\kappa})$ by Eq. (1.35):

$$\Phi_\varepsilon(\vec{\kappa}) = \frac{0.063\sigma_\varepsilon^2 L_z L_\|^2}{\left(1+\kappa_\|^2 L_\|^2 + \kappa_z^2 L_z^2\right)^{11/6}}, \qquad (6.60)$$

where $\kappa_\| = \sqrt{\kappa_x^2 + \kappa_y^2}$. This spectrum takes into account the finite external scales along the vertical (L_z) axis and horizontal plane $(L_\|)$ and the anisotropy of the inhomogeneities $\alpha = L_z/L_\| \neq 1$.

Taking then account of the fact that for $x >> a/m\tau_n$ (for f~10 GHz, $a/m\tau_n \sim 10$ km) small-scale inhomogeneties with $\kappa_z >> k/m\tau_n$ make the main contribution to the scattering, and bearing in mind that $m\tau_n >> 1$, we obtain for the intensity of the scattered field J_s the following expression

$$J_{sd} \approx \left\langle |W_1|^2 \right\rangle = \frac{0.055\pi^3 C_\varepsilon^2 \alpha^{-5/3}}{12\tau_1^3}\left(\frac{a}{2}\right)^{1/3} k^{-1/3}\left(\frac{x_1}{a}\right)^{-8/3} \times$$
$$F(z, z_0) \exp\left(-\frac{m\xi}{a}\tau_1\sqrt{3}\right) \qquad (6.61)$$

where $\xi = \sqrt{2a}\left(\sqrt{z} + \sqrt{z_0}\right)$. We have taken into account the contribution of the first mode with $n = 1$, the scattering of the other modes can be estimated similarly, however, the contribution of the modes with higher indices n decays as n^{-2}.

The contribution of the large-scale inhomogeneities to the intensity of the scattered field is exponentially small which can be explained as follows: The wave diffracted over the sphere's surface creates the secondary waves sliding along the tangent to that surface. These waves then scatter on the imhomogeneities of the refractive index at the scattering angle $\vartheta \sim \lambda/l_z$. The scattered wave touches the spherical surface at the distance from the point of initial detachment $\Delta x \sim a\vartheta = a\,\kappa_z/k$, and thereafter diffracts along the earth's surface arriving at the receiving point. Along the path x_1 derived along the geodesic curve, the wave attenuation is determined by $\exp\left(-mx_1/a\,\tau_1\sqrt{3}\right)$. For the scattered wave the distance along the geodesic x_1 is given by $x_1 = x - \Delta x$, since at the interval Δx the wave propagates under free-space conditions, as shown in Figure 6.5. As observed, for $\kappa_z < k/m\tau_n$, $\Delta x < am/\tau_n$, hence the larger the inhomogeneities participating in wave scattering, the shorter the "free-space" interval and, therefore, the larger the attenuation along the geodesic interval. When $x >> am/\tau_n$, the contribution of the large-scale inhomogeneties to single scattering, with $\kappa_z < k/m\tau_n$, can be neglected. With greater κ_z the scattering takes place in the higher layers of the troposphere at larger angles ϑ, thus increasing the "free-space" interval and decreasing the attenuation of the scattered wave.

In contrast to Eqs. (6.39) to (6.41), the dependence of the signal strength on the altitude z enters not only through the scattering angle $\vartheta = x_1/a$ but also through the height-gain function $\chi_n(z)$. In the range of altitudes $z << x^2/2a$ and distances $x(a/m)\sqrt{\tau_n/2}$, the scattering angle ϑ can be assumed to be independent of altitude, i.e. $\vartheta \sim x/a$; then the dependence of J_{sd} on z is determined by the factor $\exp\left[-(m/a)\sqrt{2az}\,\tau_n\sqrt{3}\right] \cdot |\chi_n|^2$. Substituting the asymptotes of $\chi_n(z)$ into Eq. (6.60) and assuming $z_0 >> m\,\tau_n/2k$, we obtain

$$J_{sd} \sim z^2\lambda^{2/3}, \text{ with } z << \frac{m}{k\sqrt{\tau_n}}, \tag{6.62}$$

$$J_{sd} \sim z^{-1/2}\lambda, \text{ with } z >> \frac{m\,\tau_n}{2k},$$

and with $z = m\tau_n/2k$ function $J_{sd}(z, \lambda)$ has a maximum in which the value of $|\chi_n(z)|$ is of the order of one.

We note that in the majority of experiments the wavelength dependence of the scattered field is proportional to the wavelength, $J_{sd}(\lambda) \sim \lambda^m$, where $0.7 < m < 1.4$ [5].

Let us consider the case of elevated antennas when $z, z_0 > m\tau_1/2k$ and compare the intensity of the scattered diffracted field J_{sd} with the intensity J_s calculated from formula (6.39) with the spectrum defined by expression (6.60). Extracting from Eq. (6.61) the effective scattering cross-section σ_d with anisotropy factor accounted for, i.e. $\sigma_d = 0.052 C_\varepsilon^2 k^{1/3} \alpha^{-8/3} (2\sin\vartheta/2)^{-11/3}$, we can represent the intensity of the scattered diffraction field in a form similar to Eq. (6.39)

$$J_{sd} = 16\frac{\sigma_d V_d}{x^2}, \tag{6.63}$$

where V_d is the effective scattering volume, which is defined by the equation

$$V_d = 0.0087\frac{\pi^3 x^3 \alpha}{16 m\tau_1^4 k\sqrt{z\cdot z_0}}. \tag{6.64}$$

Without even numerical calculation we may conclude that the contribution of the diffracted field will be an order of magnitude less than the contribution of the scattering in a "free-space" volume, defined by formula (6.39) according to the Booker–Gordon theory. The value of the theory developed in this section is that it provides the correct frequency and height dependence of the scattered field, compared with the "free-space" single scattering theory.

The correct estimation of the scattered field should require calculation of the additional terms in the scattered field not accounted for here. We may notice that given practical antennas we always have two terms in the incident field : direct wave + reflected from the ground wave(comprising the line-of-sight mechanism) and diffracted field. For simplicity, the reflected wave is omitted in the Booker–Gordon single-scattering theory, and the contribution from the transition (between line-of-sight and shadow region) is also omitted. The intensity of the total scattered field should contain four terms:

$$J_{total} = \frac{I_{total}}{I_{FS}} = 16\sigma_0 \frac{V_{fs}}{x^2} + 16\sigma_d \frac{V_d}{x^2} + 16\sigma_{d,0} \frac{V_{d,fs}}{x^2} 16\sigma_{0,d} \frac{V_{fs,d}}{x^2} \tag{6.65}$$

where the first term represents the contribution from the scattering in a free-space volume V_{fs}, as given by Eq. (6.39) ($V_{fs} \equiv V$), the second term is scattering of the diffracted field given by Eq. (6.64) and two last terms represent scattering of the line-of-sight waves into the diffracted field in the volume $V_{fs,d}$. (Figure 6.6) and the diffracted field into the free-space waves in the volume $V_{d,fs}$. A combination of these four terms may provide observable levels of the scattered field as well as frequency and height dependence in agreement with experiment.

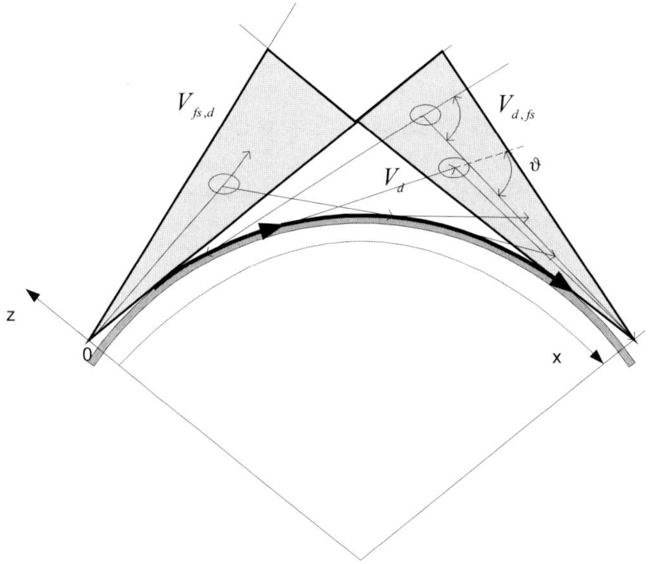

Figure 6.6 Schematic diagram of the scattering volumes for diffracted-to-diffracted field (V_d), diffracted to free-space field ($V_{d,fs}$) and free-space to diffracted field ($V_{fs,d}$).

References

1 Ivanov, V., Kinber, B., Korzenevitch, I. and Stepanov, B. Impact of the earth surface on long-range tropospheric propagation, *Radiotech. Electron.*, 1980, 25 (10), 2033–2042.

2 Saitchev, A., Slavinsky, M. Equations for moment-functions of waves propagating in random media with anisotropic inhomogeneities, *Radiophys. Quant. Electron.*, 1985, 28 (1), 75–83.

3 Kukushkin, A., Fuks, I. and Freilikher, V. Impact from random stratification on a coherent component of the field over horizon, *Radiophys. Quant. Electron.*, 1983, 28 (8), 1064–1072.

4 *Long-range UHF Propagation in the Troposphere*, Eds. Vvedensky, B., Kolosov, M., Kalinin, A. and Shifrin, J., Soviet Radio, Moscow, 1965, 416 pp.

5 Shur, A. *Signal Characteristics of Tropospheric Radiolinks*, Swyaz, Moscow, 1972, 105 pp.

6 Shifrin, J. *Problems of Statistical Antenna Theory*, Soviet Radio, Moscow, 1970, 384 pp.

7 Kalinin, A., Troitzky, V. and Shur, A. The study of long-range UHF tropospheric propagation, in *Radiowave Propagation* , Eds. Kolosov, M., Armand, N., Katzelenbaum, B. and Sokolov, A., Nauka, Moscow, 1975, pp. 127–153.

8 Rytov, S.M., Kravtsov, Y.A. and Tatarskii, V.I. *Introduction to Statistical Radiophysics: Part 2, Random Fields*, Nauka, Moscow, 1978, 464 pp.

9 Tatarskii, V.I. *The Effects of Turbulent Atmosphere on Wave Propagation*, IPST, Jerusalem, 1971.

10 Feinberg, E. *Radiowave Propagation over the Earth's Surface*, Nauka, Moscow, 1961, 547 pp.

11 Booker, H., Gordon, W. A theory of radio scattering in the troposphere, 1950, *Proc. IRE*, 1950, 38 (4), 401–412.

12 Klatskin, V.I. *Stochastic Equations and Waves in Random Media*, Nauka, Moscow, 1980, 336 pp.

13 Rotheram, S. Radiowave propagation in the evaporation duct, 1974, *The Marconi Rev.*, 1974, 37 (192), 18–40.

Appendix: Airy Functions

A.1 Definitions

Studying the behavior of light in the neighbourhood of caustics in 1838, Sir G. B. Airy introduced the famous integral

$$v(t) = \frac{1}{\sqrt{\pi}} \int_0^\infty \cos\left(\frac{x^3}{3} + xt\right) dx \tag{A.1}$$

which represents one of the solutions to the differential equation

$$w''(t) - tw(t) = 0, \tag{A.2}$$

namely, one which decreases at positive infinity more rapidly than any finite power of t. A second independent solution to Eq. (A.2) can be named $u(t)$ and will be defined later.

Following Fock [1], the solution to Eq. (A.2) can also be defined via the integral:

$$w(t) = \frac{1}{\sqrt{\pi}} \int_{\Gamma_1} dz \exp\left(tz - \frac{z^3}{3}\right) \tag{A.3}$$

where contour Γ_1 goes by the ray with $\arg z = -2\pi/3$ from infinity to 0 and then to infinity along the real and positive z-axis, Figure A.1.

Consider the behavior of the function $w(t)$ with real argument t. First, let us find the sector of convergence of the integral (A.3) in order to perform useful deformations on the integration contour. With $|z| \to \infty$ we may neglect the first term in the integrand (A.3) then the convergence sector is determined by the condition

$$\operatorname{Re} z^3 > 0. \tag{A.4}$$

We may define $z = |z|e^{j\alpha}$ $(-\pi < \alpha < \pi)$, then the condition (A.4) results in

$$\cos(3\alpha) > 0, \tag{A.5}$$

Radio Wave. Alexander Kukushkin

ISBN: 3-527-40458-9

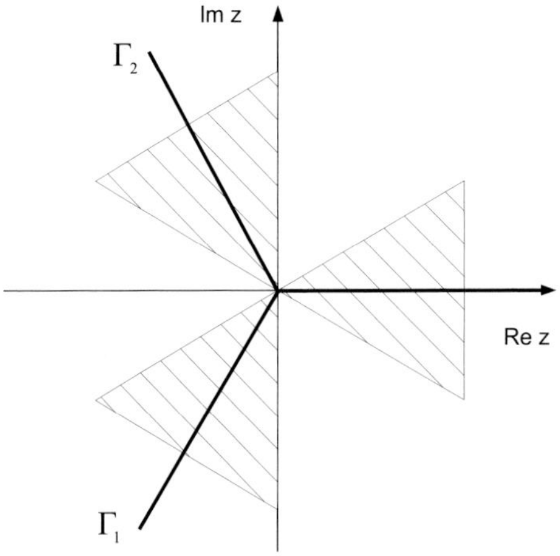

Figure A.1 Sectors of convergence for the Airy integral.

which, in turn, determines three sectors of convergence of the integral (A.3) in the z-plane:

$$1)\quad -\frac{\pi}{6} < \alpha < \frac{\pi}{6},$$

$$2)\quad \frac{\pi}{2} < \alpha < \frac{5\pi}{6}, \qquad \text{(A.6)}$$

$$3)\quad -\frac{5\pi}{6} < \alpha < -\frac{\pi}{2}.$$

The contour Γ_1 in Eq. (A.3) passes through the middle of the first and the third sector and therefore coincides with one of the lines of steepest descent of the integrand's phase.

We can now deform the contour Γ_1 within the sector of convergence without changing the value of the integral. Let us turn the lower part of contour Γ_1 to coincide with the negative imaginary axis of z. Then integral (A.3) can be written in the following form

$$w(t) = \frac{1}{\sqrt{\pi}} \int_0^{\infty} dz \exp\left(tz - \frac{z^3}{3}\right) + \frac{1}{\sqrt{\pi}} \int_{-j\infty}^{0} dz \exp\left(tz - \frac{z^3}{3}\right). \qquad \text{(A.7)}$$

Changing the variable in the second integral (A.7) to $\varsigma = jz$, we obtain

$$w(t) = \frac{1}{\sqrt{\pi}} \int_0^{\infty} dz\{\exp\left(tz - \frac{z^3}{3}\right) + \frac{j}{\sqrt{\pi}} \int_0^{\infty} d\varsigma \exp\left(-j\left(t\varsigma - \frac{\varsigma^3}{3}\right)\right). \qquad \text{(A.8)}$$

We may now select the real and imaginary parts in $w(t)$:

$$w(t) = u(t) + jv(t). \qquad \text{(A.9)}$$

where

$$u(t) = \frac{1}{\sqrt{\pi}} \int_0^\infty dz \exp\left(tz - \frac{z^3}{3}\right) + \frac{1}{\sqrt{\pi}} \int_0^\infty dz \sin\left(tz + \frac{z^3}{3}\right), \tag{A.10}$$

$$v(t) = \frac{1}{\sqrt{\pi}} \int_0^\infty dz \cos\left(tz + \frac{z^3}{3}\right) = \frac{1}{2\sqrt{\pi}} \int_0^\infty dz \exp\left(-j\left(tz + \frac{z^3}{3}\right)\right). \tag{A.11}$$

As observed from Eqs. (A.9) to (A.11), functions $u(t)$ and $v(t)$ represent two independent solutions to Eq. (A.2), and $v(t)$ is indeed an original Airy integral (A.1). The functions $u(t)$ and $v(t)$ are related via the following equation

$$u'(t)v(t) - u(t)v'(t) = 1. \tag{A.12}$$

The set of two independent solutions to Eq. (A.2) can also be found in terms of another contour integral of Eq. (A.2). Further we redefine the first integral (A.3) as a function $w_1(t)$ while the second integral we define in the following form

$$w(t) \equiv w_2(t) = \frac{1}{\sqrt{\pi}} \int_{\Gamma_2} dz \exp\left(tz - \frac{z^3}{3}\right). \tag{A.13}$$

The contour Γ_2 is a mirror image of the contour of Γ_1 relative to the real axis of z shown by the dashed line in Figure A.1. With real values of the arguments, functions $w_1(t)$ and $w_2(t)$ are complex conjugate and

$$w_2(t) = u(t) - jv(t). \tag{A.14}$$

For functions $w_2(t)$ and $w_1(t)$ there exists an equation similar to Eq. (A.10)

$$w_1'(t)w_2(t) - w_2'(t)w_1(t) = -2j \tag{A.15}$$

and

$$w_1\left(te^{j2\pi/3}\right) = e^{j\pi/3} w_2(t). \tag{A.16}$$

While functions $u(t)$ and $v(t)$ are real functions with real values of the argument t, these are also natural transcendent functions valid for complex values of t. The relations (A.9) and (A.14) hold for complex t. Numerous formulas, useful for the treatment of Airy functions, have been provided by Fock [1] and some of them are listed below for convenience:

$$w_1\left(te^{j\pi/3}\right) = 2e^{j\pi/6} v(-t), \tag{A.17}$$

$$w_1\left(te^{j\pi}\right) = u(-t) + jv(-t), \tag{A.18}$$

$$w_1\left(te^{j4\pi/3}\right) = 2e^{j\pi/6} v(t), \tag{A.19}$$

$$w_1\left(te^{j5\pi/3}\right) = e^{j\pi/3}[u(-t) - jv(-t)]. \tag{A.20}$$

The above relations represent the values of the function $w_1(t)$ at the rays $\arg t = n\pi/3$ ($n = 0, 1, 2, 3, 4, 5$) in the complex t-plane via real functions $u(t)$ and $v(t)$ with real values of the argument t.

We may also note that the functions $u(t)$ and $v(t)$ are equivalent to Airy functions $Ai(t)$ and $Bi(t)$ [2]:

$$Ai(t) = \frac{1}{\sqrt{\pi}}v(t); \quad Bi(t) = \frac{1}{\sqrt{\pi}}u(t).$$

A.2 Asymptotic Formulas for Large Arguments

The asymptotic expressions for the Airy function for large arguments have been derived by Fock and are provided below.

Frequently used formulas for large negative t are:

$$w_1(t) = (-t)^{-1/4} \exp\left[j\frac{2}{3}(-t)^{3/2} + j\frac{\pi}{4}\right], \tag{A.21}$$

$$w_2(t) = (-t)^{-1/4} \exp\left[-j\frac{2}{3}(-t)^{3/2} - j\frac{\pi}{4}\right]. \tag{A.22}$$

Let us introduce the definition

$$x = \frac{2}{3}t^{3/2}$$

and coefficients a_n, b_n to be used in the forthcoming formulas:

$$a_1 = \frac{5}{72}; \; a_2 = \frac{(5{\cdot}11){\cdot}7}{1{\cdot}2{\cdot}(72)^2}; \; a_3 = \frac{(5{\cdot}11{\cdot}17){\cdot}(7{\cdot}13)}{1{\cdot}2{\cdot}3{\cdot}(72)^3};$$

$$a_n = \frac{(5{\cdot}11 \ldots (6n-1)){\cdot}(7{\cdot}13 \ldots (6n-5))}{1{\cdot}2{\cdot}\ldots n(72)^n},$$

$$b_1 = \frac{7}{72}; \; b_2 = \frac{(7{\cdot}13){\cdot}5}{1{\cdot}2{\cdot}(72)^2}; \; b_3 = \frac{(7{\cdot}13{\cdot}19){\cdot}(5{\cdot}11)}{1{\cdot}2{\cdot}3{\cdot}(72)^3};$$

$$b_n = \frac{(7{\cdot}13 \ldots (6n+1)){\cdot}(5{\cdot}11 \ldots (6n-7))}{1{\cdot}2{\cdot}\ldots n(72)^n}.$$

Then a comprehensive asymptotic expression for Airy functions of large positive argument t can be written as follows:

$$u(t) = t^{-1/4}e^x\left(1 + \frac{a_1}{x} + \frac{a_2}{x^2} + \ldots\right), \tag{A.23}$$

$$u'(t) = t^{1/4} e^{x} \left(1 - \frac{b_1}{x} - \frac{b_2}{x^2} - \ldots\right), \tag{A.24}$$

$$v(t) = \frac{1}{2} t^{-1/4} e^{-x} \left(1 - \frac{a_1}{x} + \frac{a_2}{x^2} - \frac{a_3}{x^3} + \ldots\right), \tag{A.25}$$

$$v'(t) = -\frac{1}{2} t^{1/4} e^{-x} \left(1 + \frac{b_1}{x} - \frac{b_2}{x^2} + \frac{b_3}{x^3} - \ldots\right). \tag{A.26}$$

For large negative t we have

$$u(-t) = t^{-1/4} \cos\left(x + \frac{\pi}{4}\right)\left(1 - \frac{a_2}{x^2} + \frac{a_4}{x^4} - \frac{a_6}{x^6} + \ldots\right) + t^{-1/4} \sin\left(x + \frac{\pi}{4}\right)\left(\frac{a_1}{x} - \frac{a_3}{x^3} + \frac{a_5}{x^5} - \frac{a_7}{x^7} + \ldots\right), \tag{A.27}$$

$$u'(-t) = t^{1/4} \sin\left(x + \frac{\pi}{4}\right)\left(1 + \frac{b_2}{x^2} - \frac{b_4}{x^4} + \frac{b_6}{x^6} - \ldots\right) + t^{1/4} \cos\left(x + \frac{\pi}{4}\right)\left(\frac{b_1}{x} - \frac{b_3}{x^3} + \frac{b_5}{x^5} - \frac{b_7}{x^7} + \ldots\right), \tag{A.28}$$

$$v(-t) = t^{-1/4} \sin\left(x + \frac{\pi}{4}\right)\left(1 - \frac{a_2}{x^2} + \frac{a_4}{x^4} - \frac{a_6}{x^6} + \ldots\right) - t^{-1/4} \cos\left(x + \frac{\pi}{4}\right)\left(\frac{a_1}{x} - \frac{a_3}{x^3} + \frac{a_5}{x^5} - \frac{a_7}{x^7} + \ldots\right), \tag{A.29}$$

$$v'(-t) = t^{1/4} \cos\left(x + \frac{\pi}{4}\right)\left(1 + \frac{b_2}{x^2} - \frac{b_4}{x^4} + \frac{b_6}{x^6} - \ldots\right) + t^{1/4} \sin\left(x + \frac{\pi}{4}\right)\left(\frac{b_1}{x} - \frac{b_3}{x^3} + \frac{b_5}{x^5} - \frac{b_7}{x^7} + \ldots\right). \tag{A.30}$$

A.3 Integrals Containing Airy Functions in Problems of Diffraction and Scattering of UHF Waves

In this section we obtain the asymptotical expansions of the integrals containing the product of the Airy–Fock functions [1]:

$$V = \int_{-\infty}^{0} w(t - y) w(t - y_0) \exp(j\xi t) dt \tag{A.31}$$

where $w(x)$ is any one of the solutions to the Airy equation

$$w'(x) - x w(x) = 0 \tag{A.32}$$

defined by one of the integrals below with desirable behavior at infinity

$$w(x) \equiv w_1(x) = \frac{1}{\sqrt{\pi}} \int_\Gamma dz \exp\left(xz - \frac{z^3}{3}\right), \tag{A.33}$$

$$w_1(x)|_{x\to\infty} \sim (-x)^{-1/4} \exp\left[j\frac{2}{3}(-x)^{3/2} + j\frac{\pi}{4}\right], \tag{A.34}$$

$$w(x) \equiv v(x) = \frac{1}{2\sqrt{\pi}} \int_{-\infty}^{\infty} dz \exp\left[j\frac{z^3}{3} + jzx\right], \tag{A.35}$$

$$v(x)|_{x\to\infty} \sim (x)^{-1/4} \exp\left[-\frac{2}{3}(x)^{3/2}\right]. \tag{A.36}$$

In integral (A.33) the contour Γ elapses from ∞ along the ray $\exp(j\,4\pi/3)$ towards 0 and then along the real axis of z to $+\infty$. Figure A.1. Along with the functions $w_1(x)$, $v(x)$ we use function $w_2(x) = w_1^*(x)$, where the sign * denotes complex conjugate.

Integrals of type (A.31) are often calculated in problems of UHF diffraction and propagation in the atmospheric boundary layer. This is because in an analytical approach the model of the average refractivity can be described in terms of the linear approximation to a whole height profile of the averaged refractivity or, at least, to some sections of the profile. For instance, the integral of type (A.31) appears when one need coefficients of the energy transformation between modes (coupling coefficients) due to scattering on the random fluctuations of the refractive index in the medium. Then the eigenfunctions of the operator

$$L_z[\varphi_z] = \left[\frac{d^2}{dz^2} - k^2(\varepsilon_0 - g_\varepsilon z)\right]\varphi_n(z) = E_n\varphi_n(z), \tag{A.37}$$

$$\varphi_n(z=0) = 0, \quad \varphi_n(z\to\infty) = 0 \tag{A.38}$$

associated with the discrete spectrum of eigenvalues correspond to the waves propagating with almost negligible attenuation, i.e. waveguide modes (trapped modes in the common terminology of tropospheric duct propagation). In the case of linear profile $\varepsilon(z)$these functions can be expressed via the Airy function

$$\varphi_n(z) = C_n v\left(\frac{E_n - \mu^3(z-H)}{\mu^2}\right) \tag{A.39}$$

where z is the height above the boundary surface $(z = 0)$, $\mu^3 = ag_\varepsilon/2$ the gradient of the refractive index inside the waveguide, i.e. for $0 < z < H$, H the thickness of the waveguide formed by a negative gradient of the refractive index g_ε, $g_\varepsilon = -(d\varepsilon/dz + 2/a)$.

The coefficient of re-scattering $V_{m,n}$ between the modes with numbers m and n due to scattering on a spatial Fourier-component of the fluctuation in the refractive index with vertical wavenumber κ is then given by

$$V_{m,n}(\kappa) = \int_0^\infty dz v\left(\frac{E_n - \mu^3(z-H)}{\mu^2}\right) v\left(\frac{E_m - \mu^3(z-H)}{\mu^2}\right) \exp(j\kappa z). \tag{A.40}$$

After apparent transformations, the evaluation of integral (A.40) will result in calculation of the integral

$$V_1 = \int_0^\infty dt \cdot v(t - \xi_n) v(t - \xi_m) \exp(jqt). \tag{A.41}$$

Similar arguments can be applied to evaluation of the coupling coefficients of the eigenfunctions of the continuous spectrum $\Psi_{E_1}(z)$ and $\Psi_{E_2}(z)$. These coefficients will result in integrals of the following kind:

$$V_2 = \int_{-\infty}^0 dt \cdot w_1(t - \xi_1) w_1(t - \xi_2) \exp(jqt), \tag{A.42}$$

$$V_3 = \int_{-\infty}^0 dt \cdot w_1(t - \xi_1) v(t - \xi_2) \exp(jqt), \tag{A.43}$$

$$V_4 = \int_{-\infty}^0 dt \cdot w_1(t - \xi_1) w_2(t - \xi_2) \exp(jqt) = J_1 - 2jJ_3. \tag{A.44}$$

We can note that

$$V_5 = \int_{-\infty}^0 dt \cdot w_2(t - \xi_1) w_2(t - \xi_2) \exp(jqt) = J_2(-q, \xi_1, \xi_2) \tag{A.45}$$

for real q, ξ_1 and ξ_2. Therefore, the basic set of integrals of interest is J_1, J_2 and J_3.

A.3.1
Integral V_1

Making use of the integral representation (A.35), transform integral V_1 into a triple integral and perform integration over the variable t as well as over one of the newly introduced variables of integration. As a result we obtain

$$V_1 = \frac{\exp\left(j^{\pi/4}\right)}{4\sqrt{\pi}} \int_{-\infty}^{\infty} \frac{ds}{\sqrt{s(s+q)}} \exp\left(j\frac{s^3}{12} - jsv - j\frac{\delta}{4s}\right) \tag{A.46}$$

where the contour of integration over s encircles the singularities $s = 0$ and $s = q$ in the upper half-plane of the variable s; $\nu = (\xi_1 + \xi_2)/2$, $\delta = \xi_1 - \xi_2$. With $|\nu| >> 1$ integral (A.46) can be calculated using the method of stationary phase.

Assume $\nu >> 1$, $\delta >> 1$. In the integral (A.1.1) we observe four stationary points $s_{1,2,3,4} = \pm(\sqrt{\xi_1} \pm \sqrt{\xi_2})$. The pair $s_{3,4} = \pm(\sqrt{\xi_1} - \sqrt{\xi_2}) \approx \pm\dfrac{\delta}{2\sqrt{\nu}}$ may come close to the square root singularity in the point $s = 0$.

Let us evaluate the contribution of the stationary points $s_{1,2} = \pm(\sqrt{\xi_1} + \sqrt{\xi_2}) \approx 2\sqrt{\nu}\left(1 - \dfrac{\delta^2}{32\nu^2}\right)$. We may notice that

$$V_1 = \frac{\exp\left(-j^{\pi/4}\right)}{4\sqrt{\pi}}\left(I_1(\nu,\delta,q) + jI_1^*(\nu,\delta,-q)\right), \tag{A.47}$$

where

$$I_1 = \int_0^\infty \frac{ds}{\sqrt{s(s+q)}} \exp\left(j\frac{s^3}{12} - js\nu - j\frac{\delta}{4s}\right). \tag{A.48}$$

With $\frac{\delta}{2\sqrt{\nu}} << 1$

$$I_1(\nu,\delta,q) \cong I_1(\nu,0,q) = \frac{\exp\left(j^{4/3}\nu^{3/2}\right)}{\sqrt{2\sqrt{\nu}}} \int_{-\infty}^{\infty} \frac{dx}{(x+2\sqrt{\nu}+q)} \exp\left(j\frac{s^3}{2} - js\nu - j\frac{\delta}{4s}\right) \tag{A.49}$$

Taking into account the known representation for the error integral [2], we obtain

$$I_1(\nu,\delta,q) \cong -j\frac{\pi}{\sqrt{2\sqrt{\nu}}}\left\{1 + \frac{2}{\sqrt{\pi}}\exp\left(j\frac{\pi}{4}\right)\int_{-\infty}^{\nu^{1/4}(s_1+q)/\sqrt{2}} dx \exp\left(-jx^2\right)\right\} \tag{A.50}$$

and, respectively, a uniform asymptotic for the contribution to J_1 from stationary points $s_{1,2} = \pm 2\sqrt{\nu}$:

$$V_1 \cong j\frac{\pi}{\sqrt{2\sqrt{\nu}}}\left\{\begin{array}{l} \exp\left(-j\frac{4}{3}\nu^{3/2} + j\frac{\sqrt{\nu}}{2}\left(s_{1,2} + q^2\right)\right)\times \\ \left[\frac{1+2e^{j\pi/4}}{\sqrt{\pi}}\int_{-\infty}^{\nu^{1/4}(s_{1,2}+q)/\sqrt{2}} \exp\left(-jx^2\right)dx\right] - \\ j\exp\left(-j\frac{4}{3}\nu^{3/2} - j\frac{\sqrt{\nu}}{2}\left(s_{1,2} + q^2\right)\right)\times \\ \left[1 + \frac{2e^{-j\pi/4}}{\sqrt{\pi}}\int_{-\infty}^{\nu^{1/4}(s_{1,2}+q)/\sqrt{2}} \sqrt{2}\exp\left(jx^2\right)dx\right] \end{array}\right\}. \tag{A.51}$$

Now, let us evaluate the contribution to V_1 of the region in the vicinity of the stationary points $s_{3,4} \approx \pm\delta/2\sqrt{\nu}$. We may notice that with $\delta >> 2\sqrt{\nu}$ evaluation of the integral can be performed similarly to the method used in obtaining Eq. (A.51). Therefore, attention will be paid to the case $\delta << 2\sqrt{\nu}$, which, being applied to the problem of scattering on random fluctuations in the refractive index, corresponds to the situation where the transversal wavenumbers of the mode's pair are large, though the distance between the modes in a wavenumber space is small. This situation occurs in a multimode tropospheric duct for higher-order modes when the duct is formed, for instance, by inversion of temperature.

Taking into account that the major contribution to the integral comes from the region of small s, $s \sim 1/\nu$, we may expand $\exp(js^3/12)$ into a Taylor series:

$$V_1 \cong -\frac{e^{-j\pi/4}}{4\sqrt{\pi}} \int_{-\infty}^{\infty} ds \frac{\exp\left(-js\nu - j\frac{\delta^2}{4s}\right)}{\sqrt{s(s+q)}} \left(1 + j\frac{s^3}{12} - j\frac{s^9}{288} + \ldots\right)$$
$$= \frac{e^{j\pi/4}}{4\sqrt{\pi}} \left[I_2(\nu, \delta, q) + \frac{1}{12}\frac{\partial^3}{\partial \nu^3} I_2(\nu, \delta, q) + \ldots \right]. \tag{A.52}$$

We may observe that

$$I_2 = e^{j\nu q} I_3; \qquad I_3 = \int_{-\infty}^{\infty} ds \frac{\exp\left(-j\nu(s+q) - j\frac{\delta^2}{4s}\right)}{\sqrt{s(s+q)}}. \tag{A.53}$$

Now, introduce a new function I_4 using the relation

$$I_4 = \frac{\partial I_3}{\partial \nu} j e^{j\nu q}. \tag{A.54}$$

The function I_4 is then determined via the integral

$$I_4 = \int_0^{\infty} \frac{ds}{\sqrt{s}} \left[\cos(\psi(s, \nu, \delta)) + \sin(\psi(s, \nu, \delta))\right] \tag{A.55}$$

where $\psi(s, \nu, \delta) = \nu s + \delta/4s$. The integral (A.55) is tabulated [3] and given by

$$I_4 = \sqrt{\frac{2\pi}{\sqrt{\nu}}} \cos(\delta\sqrt{\nu}).$$

Then, the reverse operation leads to

$$I_2 = 2\sqrt{\pi} e^{-j3\pi/4 + j\nu q} \int_{\infty}^{\nu} d\eta e^{j\eta q} \frac{\cos(\delta\sqrt{\eta})}{\sqrt{\eta}} + C. \tag{A.56}$$

The value of constant C, which does not depend on parameter q, can be chosen using the norm of J_1 with $q \to 0$. Then,

$$C = j\frac{2\pi}{\sqrt{q}} \exp\left(j\frac{\delta^2}{4q} - j\frac{\pi}{4}\right).$$

After some transformation expression for I_2 the following results:

$$I_2(\nu, \delta, q) = -\frac{2\sqrt{\pi}}{\sqrt{q}} \exp\left(-j\frac{3\pi}{4} + j\nu q + j\frac{\delta^2}{4q}\right) \times$$
$$\left\{ \mathrm{erf}\left[e^{j\pi/4}\left(\sqrt{q}\nu - \frac{\delta}{2\sqrt{q}}\right)\right] + \mathrm{erf}\left[e^{j\pi/4}\left(\sqrt{q}\nu + \frac{\delta}{2\sqrt{q}}\right)\right] \right\} \tag{A.57}$$

where $\mathrm{erf}(.)$ is the error integral [2].

It should be noted that the second term in Eq. (A.52) has an order $O(1/\nu^{3/2})$ and may be neglected for the following reason. We seek the estimate of the integral J_1 as a sum of Eqs. contributions from the stationary points, four in total. Therefore, the result is a combination of (A.51) and (A.52). Taking into account that the contribution of the stationary points $s_{1,2} = \pm 2\sqrt{\nu}$ in Eq. (A.51) evaluated with precision up to the terms of order $O(1/\sqrt{\nu})$, we should neglect the second term in Eq. (A.52), thus substituting the expression (A.57) for I_2 in Eq. (A.52).

Finally with $\nu >> 1$, $\delta << 1$, the integral $V_1(\nu, \delta, q)$ is represented by a sum of contributions of four stationary points:

$$V_1(\nu, \delta, q) \approx V_1|_{s=s_1} + V_1|_{s=s_{21}} + V_1|_{s=s_3} + V_1|_{s=s_4},$$

where $V_1|_{s=s_1}$ and $V_1|_{s=s_2}$ are defined by Eq. (A.51) and $V_1|_{s=s_3}$ and $V_1|_{s=s_4}$ by Eqs. (A.52) and (A.57).

In the limiting case $\delta = 0$ and $q\nu << 1$ we obtain

$$V_1(\nu, \delta, q) = \sum_{p=0}^{\infty} \frac{2^p e^{j\pi p/2}}{(2p+1)} q^p \nu^{p+1/2} + O\left(\frac{1}{\sqrt{\nu}}\right). \tag{A.58}$$

With $q = 0$, $\nu = \xi_n$ (recall that ξ_n is a propagation constant of the nth mode), the expression (A.58) equates to the known asymptotic of the norm of the eigenfunctions of the discrete spectrum with $\xi_n >> 1$:

$$N_n = \int_0^{\infty} dt v^2 (t - \xi_n) = \sqrt{\nu}. \tag{A.59}$$

a) Let us consider the integral J_1 with small parameters δ, ν: $\nu << 1$, $\delta << 1$. Expanding $\exp(-j\nu s - j\delta^2/4s)$ and $(s+q)^{-1}$ into a series, we obtain

$$V_1(\nu, \delta, q) = \sum_{m=0}^{\infty} \sum_{p=0}^{\infty} \sum_{r=0}^{\infty} B_{pmr} \frac{(-1)^r e^{-j\pi p + m/2}}{p!m!2^{2m}} q^r \nu^p \tag{A.60}$$

where

$$B_{pmr} = (12)^{\frac{m-p+r}{3}} \Gamma\left(\frac{p-m+r}{3} - \frac{1}{6}\right) \times$$
$$\left[\exp\left(j\frac{\pi}{6}\left(p - m - r - \frac{1}{2}\right)\right) + \exp\left(j\frac{5\pi}{6}\left(p - m - r - \frac{1}{2}\right)\right)\right].$$

In the particular case when $q = \nu = \delta = 0$, integral (A.1.15) equates to

$$V_1 = \frac{\Gamma\left(\frac{5}{6}\right) 2^{-4/3}}{\sqrt{\pi} 3^{1/6}} \tag{A.61}$$

which is the same as the known integral [2]:

$$\int_0^{\infty} dx v^2(x) = \left[-\frac{\sqrt{\pi}}{3^{1/3}\Gamma\left(\frac{1}{3}\right)}\right]^2 = \frac{\Gamma\left(\frac{5}{6}\right) 2^{-4/3}}{\sqrt{\pi} 3^{1/6}}. \tag{A.62}$$

b) With $q\nu >> 1$ we may expand $(s+q)^{-1}$ over the reverse powers of q thus obtaining the series representation

$$V_1(\nu,\delta,q) = \sum_{m=0}^{\infty}\sum_{p=0}^{\infty}\sum_{r=0}^{\infty} B_{pmr}\frac{(-1)^r e^{-j\pi\, p+m/2}}{p!m!2^{2m}}\frac{\nu^p\delta^{2m}}{q^{r+1}} \tag{A.63}$$

where

$$B_{pmr} = (12)^{\frac{m-p+r}{3}-\frac{1}{6}}\Gamma\left(\frac{p-m+r}{3}-\frac{1}{6}\right)\times$$
$$\left[\exp\left(j\frac{\pi}{6}\left(p-m-r+\frac{1}{2}\right)\right)+\exp\left(j\frac{5\pi}{6}\left(p-m-r+\frac{1}{2}\right)\right)\right]. \tag{A.64}$$

c) When $\nu \sim 1$, $\delta << 1$, we may expand $\exp(js\nu + j\,\delta/4s)$ over a series of Bessel functions, thus obtaining

$$V_1(\nu,\delta,q) = \sum_{m=-\infty}^{\infty}(-j)^m J_m\left(\delta\frac{\sqrt{\nu}}{2}\right)\left(\frac{2\sqrt{\nu}}{\delta}\right)^m B_m(q) \tag{A.65}$$

where $J_m(x)$ is a Bessel function of the first kind and coefficient $B_m(q)$ is given by the integral

$$B_m(q) = \int_{-\infty}^{+\infty} ds\frac{\exp\left(j\frac{s^3}{12}\right)}{s+q}s^{m-\frac{1}{2}}, \tag{A.66}$$

which can be represented via Gamma-functions similar to the coefficient in Eqs. (A.61) and (A.64).

A.3.2 Integral V_2

Let us consider integral (A.42) where ξ_1 and ξ_2 have real and positive values, q is a parameter of real value.

$$V_2 = \int_{-\infty}^{0} dt\cdot w_1(t-\xi_1)w_1(t-\xi_2)\exp(jqt)$$

We use the integral representation for Airy functions (A.33) to transform the integral (A.42). The contour of the integration Γ_1 in (A.33) can be defined as elapsing from $-j\infty$ to 0 and then to ∞ along the real axis. After integration over variable t we obtain

$$V_2 = \frac{1}{\pi}\int_{\Gamma}\int_{\Gamma} d\varsigma_1 d\varsigma_2\frac{\exp\left(-\xi_2\varsigma_1-\xi_1\varsigma_2-\frac{\varsigma_1^3+\varsigma_2^3}{3}\right)}{\varsigma_1+\varsigma_2+\sigma+jq}. \tag{A.67}$$

We have introduced an additional small parameter σ, $\sigma > 0$, in order to shift the pole from the contour of integration into point $-(\sigma + jq)$ and make a transformation of the contour Γ_1 into contour Γ_1' as shown in Figure A.2.

Introducing a new variable $s = \varsigma_1 + \varsigma_2$ and redefining for convenience $\varsigma \equiv \varsigma_1$, we may find that integration over variable ς can then be performed given the convergence after changing the order of integration:

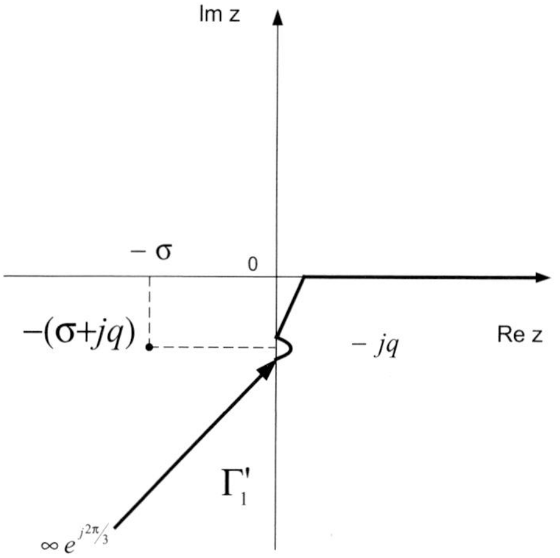

Figure A.2 Contour of integration for integral V_2.

$$V_2 = \frac{1}{\sqrt{\pi}} \int_{\Gamma_2} ds \frac{\exp\left(-\frac{s^3}{12} - s\frac{\xi_1+\xi_2}{2} + \frac{(\xi_1-\xi_2)^2}{4s}\right)}{\sqrt{s}(s+jq+\sigma)}. \tag{A.68}$$

The contour Γ_2 for the newly introduced variable s can be obtained from contour Γ_1 by shifting Γ_1 to a value ς_2. When ς_2 varies along the imaginary axis, the contour Γ_1 is shifted down along that axis to $|\varsigma_2|$. When $\varsigma_2 \geq 0$, the contour Γ_1 should be shifted right along the real axis, i.e. contour Γ_2 of the integration over s seems to depend on the value of ς_2. We may demonstrate that all these variations in contour Γ_2 are equivalent to contour Γ_1.

Consider first the case, when the contour Γ_1 is shifted along the imaginary axis s to the value $-|\varsigma_2|$ as shown in Figure A.3. Inside the closed contour in Figure A.3, integral V_2 has no singularities and along the branches C_1 and $C_2$of the contour its magnitude is negligible. Therefore, according to Cauchy's theorem, we may perform integration over contour $\Gamma_1$instead of Γ_2, i.e. these contours are equivalent in terms of integration.

Using similar arguments we may show that when the contour Γ_2 is obtained by a shift of Γ_1 along the real axis, both contours are equivalent. Therefore, whatever value ς_2 takes along contour Γ, contour Γ_2 can be deformed into contour Γ_1 which, in turn, no longer depends on value ς_2, thus allowing, in principle, the changes in the order of integration in Eq. (A.68).

In order to perform the integration we need to deform the contours Γ and Γ_2 within their sector of convergence in such a way that will ensure convergence of the integral over variable ς. New contours will have the following shape: contour Γ goes from negative ∞ along the ray $e^{-j2\pi/3}$ to 0 and then along the real axis to ∞; contour Γ_2 goes from $-j\infty$ to 0 overtaking the pole $-jq$ from the right (in a counter clockwise direction) and then to ∞ along the ray $e^{-j\pi/6}$.

The internal integral over ς is then a Poisson's integral with apparent solution. In the remaining single integral we then make a further transformation of the contour

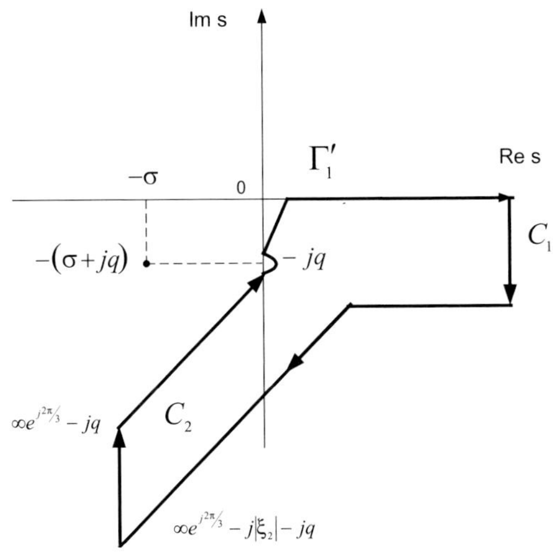

Figure A.3 Transformation of the integration contour in V_2.

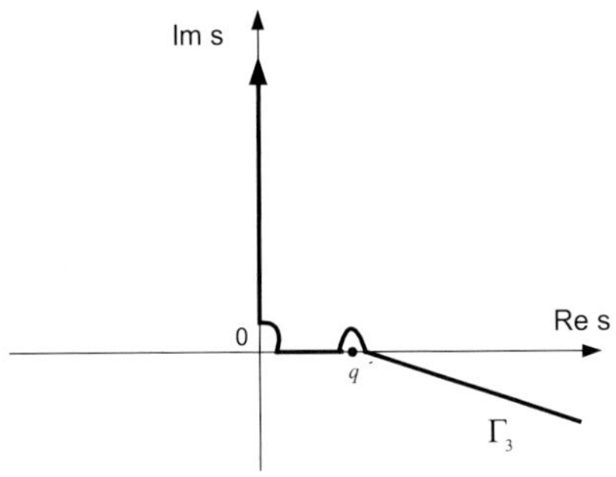

Figure A.4 Final integration contour in V_2.

of integration Γ_2, namely we can turn the contour Γ_2 through an angle of $\pi/2$ in a counterclockwise direction and change the path along the contour in the opposite direction, then parameter σ can be put to 0.

Finally, we end up with the following integral

$$V_2 = -\frac{e^{j\frac{\pi}{4}}}{\sqrt{\pi}} \int_{\Gamma_3} ds \frac{\exp\left(-j\frac{s^3}{12} + jsv - j\frac{\delta^2}{4s}\right)}{\sqrt{s}(s-q)} \tag{A.69}$$

with contour Γ_3 shown in Figure A.4.

Now we can evaluate integral (A.69) in some limiting cases.

a) Consider the case when $\nu >> 1$, $\delta << 1$.

The integrand in Eq. (A.69) along the imaginary axis represents a rapidly attenuating function without any singularities at that section of integration. The major contribution to Eq. (A.69) hence follows from the integration along the ray $e^{-j\pi/6}$ (or along the real axis of s, since that section of the contour Γ_3 can be further transformed to one along the real axis).

The asymptotic evaluation of the integral (A.69) is then composed of a contribution from the stationary points: $s_1 = 2\sqrt{\nu}\left(1 - \delta^2/32\nu\right)$ and $s_3 = \delta/2\sqrt{\nu}$:

$$V_2 \approx V_2|_{s_1} + V_2|_{s_3} \tag{A.70}$$

where

$$V_2|_{s_1} \cong -\frac{e^{j\pi/4}}{\sqrt{\pi}} I_5(\nu, \delta, q) \tag{A.71}$$

and

$$I_5(\nu, \delta, q) = \int_0^\infty ds \frac{\exp\left(-j\frac{s^3}{12} + js\nu\right)}{\sqrt{s(s-q)}}. \tag{A.72}$$

Using an approach similar to that introduced in Section A.3.1, we obtain the asymptotic for $I_5(\nu, \delta, q)$:

$$I_5(\nu, \delta, q) \cong j\pi\sqrt{\frac{2}{\sqrt{\nu}}}\exp\left(-j\frac{4}{3}\nu^{3/2} - j\frac{\sqrt{\nu}}{2}(s_1 - q)^2\right) \times \left[1 + \frac{e^{-j\pi/4}}{\sqrt{\pi}} \int_0^{\nu^{1/4}(s_1 - q)/2} \exp(j\tau^2) d\tau\right]. \tag{A.73}$$

A contribution from the stationary point s_3 into V_2 will be evaluated when the stationary point s_3 is close to a square root singularity, i.e when $\delta << 2\sqrt{\nu}$. Expanding $\exp\left(-js^3/12\right)$ into a Taylor's series in the vicinity of $s = 0$ we obtain

$$\begin{aligned} V_2|_{s_3} &\cong -\frac{e^{j\pi/4}}{4\sqrt{\pi}} \sum_{n=0}^\infty \frac{e^{-j\pi n/2}}{12^n n!} \int_{-\infty}^\infty ds \frac{\exp\left(js\nu + j\frac{\delta^2}{4s}\right)}{\sqrt{s(s-q)}} s^{3n} \\ &= -\frac{e^{j\pi/4}}{4\sqrt{\pi}}\left[I_6(\nu, \delta, q) + \frac{1}{12}\frac{\partial^3}{\partial \nu^3} I_6(\nu, \delta, q) + \ldots\right] \end{aligned} \tag{A.74}$$

Then using similar arguments as in the derivation of Eq. (A.56) we obtain for I_6 the following expression

$$I_6(\nu, \delta, q) = e^{j3\pi/4}\sqrt{\pi} \int_{-\infty}^{\nu} dx \frac{\exp\left(-jxq + j\delta\sqrt{x}\right)}{\sqrt{x}} + C. \tag{A.75}$$

The value of constant C is determined from the condition of the finite value of V_2 with $q \to 0$:

$$C = -\sqrt{\frac{\pi}{\theta}} \exp\left(j\frac{\delta^2}{4q} - j\frac{\pi}{4}\right), \tag{A.76}$$

and the asymptotic for $V_2|_{s_3}$ is finally given by

$$V_2|_{s_3} \cong 2\sqrt{\frac{\pi}{\theta}} \exp\left(j\nu q + j\frac{\delta^2}{4q}\right) \operatorname{erf}\left(e^{j\pi/4}\left(\sqrt{\nu q} + \frac{\delta}{2\sqrt{q}}\right)\right). \tag{A.77}$$

b) For small parameters ν, δ, i.e., $\nu << 1$, $\delta << 1$, we may use a series expansion for V_2 similar to the approach in Section A.3.1.

For the case $q\nu >> 1$,

$$V_2 = -\frac{e^{j\pi/4}}{\sqrt{\pi}} \sum_{m=0}^{\infty} \sum_{p=0}^{\infty} \sum_{r=0}^{\infty} B_{pmr} \frac{e^{-j\pi p + m/2 - j\pi}}{p!m!2^{2m}} \frac{\nu^p \delta^{2m}}{q^{r+1}} \tag{A.78}$$

where

$$B_{pmr} = 4 \cdot (12)^{\frac{p+r-m}{3}+\frac{5}{6}} \Gamma\left(\frac{p-m+r}{3}+\frac{1}{6}\right) \times \left[\exp\left(j\frac{\pi}{2}\left(p-m+r-\frac{3}{2}\right)\right) + \exp\left(j\frac{\pi}{6}\left(p-m-r+\frac{1}{2}\right)\right)\right]. \tag{A.79}$$

For the case $q\nu << 1$,

$$V_2 = -\frac{e^{j\pi/4}}{\sqrt{\pi}} \sum_{m=0}^{\infty} \sum_{p=0}^{\infty} \sum_{r=0}^{\infty} B_{pmr} \frac{e^{j\pi p + m/2}}{p!m!2^{2m}} \nu^p \delta^{2m} q^r \tag{A.80}$$

where

$$B_{pmr} = 4 \cdot (12)^{\frac{p+r-m}{3}-\frac{7}{6}} \Gamma\left(\frac{p-m+r}{3}-\frac{1}{6}\right) \times \left[\exp\left(j\frac{\pi}{2}\left(p-m+r-\frac{5}{2}\right)\right) + \exp\left(-j\frac{\pi}{6}\left(p-m-r-\frac{1}{2}\right)\right)\right]. \tag{A.81}$$

With $\nu \sim 1$ and $\delta \leq 1$, the asymptotic of V_2 can be obtained via a series of Bessel functions in a way similar to the approach in Section A.3.1.

A.3.3 Integral V_4

We use the integral representations (A.3), (A.11) of Airy functions $w_1(.)$ and $v(.)$ to calculate V_4. We may choose a contour of integration for $w_1(.)$ in Eq. (A.3) to pass along the negative imaginary axis to 0 and then along the real axis to ∞. The integral representation for $v(.)$ can be chosen as follows:

$$v(x) = \frac{1}{2\sqrt{\pi}} \int_{-j\infty}^{j\infty} e^{xz - \frac{z^3}{3}} dz. \tag{A.82}$$

Then we can perform integration over t in V_4 and, in a remaining double integral, we can change the variables s, ς similar to those used in Section A.2. Using similar arguments for contour transformation we truncate the integral V_4 to a single integral given by

$$V_4 = \frac{e^{j\frac{3\pi}{4}}}{2\sqrt{\pi}} \int_{-\infty}^{\infty} ds \frac{\exp\left(-j\frac{s^3}{12} + js\nu + j\frac{\delta^2}{4s}\right)}{\sqrt{s(s-q)}}. \tag{A.83}$$

The contour of integration takes into account the overpass of both the square root singularities and the pole over the upper half-circumference in a clockwise direction. The asymptotes for V_4 can be obtained in a way similar to that used for V_2 in Section A.2, taking into account the contribution of the stationary points $s_4 = -2\sqrt{\nu}$ and $s_5 = -\delta/2\sqrt{\nu}$. We may also observe that $V_4 = 2V_1^*(\nu, \delta, -q)$.

A.3.4
Integral V_5

As mentioned earlier, the integral

$$V_5 = \int_{-\infty}^{0} dt \cdot w_2(t-\xi_1) w_2(t-\xi_2) \exp(jqt) \tag{A.84}$$

can be expressed via the basic integral V_2. Here, we obtain the asymptotic form of V_5 for small values of the parameter δ, $\delta << 1$.

Let us introduce a new variable $\tau = t - \xi_2$ and transform the contour of integration in Eq. (A.84) to a ray $\infty e^{j2\pi/3}$. Taking into account known transformations of the Airy functions [1],

$$w_2\left(\tau e^{j2\pi/3}\right) = 2e^{j\pi/6} v(\tau),$$

$$w_2\left(\tau e^{j2\pi/3} - \delta\right) = 2e^{-j\pi/6} v(\tau - \delta e^{-j2\pi/3})$$

we obtain

$$V_5 = 4e^{j\pi/3 - j\xi_2 q} \int_{\xi_2}^{\infty} d\tau \cdot v(\tau) v\left(\tau - \delta e^{-j2\pi/3}\right) \exp\left(-q\tau \frac{(\sqrt{3}+j)}{2}\right). \tag{A.85}$$

Assume both $\delta << 1$, $|q| << 1$,

$$V_5 = 4e^{j\pi/3 - j\xi_2 q} \sum_{m=0}^{\infty} \sum_{n=0}^{\infty} \frac{(-1)^n \exp\left(-j\pi \frac{\left(2n + \frac{7m}{2}\right)}{3}\right)}{n!m!} q^m \delta^n \times \tag{A.86}$$

$$\int_{\xi_2}^{\infty} d\tau \cdot \tau^m v(\tau) \frac{d^n}{d\tau^n} v(\tau).$$

As observed, the task is truncated to calculation of the integrals of the following type:

$$Q_{m,n} = \int \tau^m v(\tau) \frac{d^n}{d\tau^n} v(\tau) d\tau. \tag{A.87}$$

In the case when $n = 0$, $m \geq 3$, integrals $Q_{m,0}$ can be easily calculated using the recurrent formula:

$$Q_{m,0} = \frac{m}{2m+1}\left[\begin{array}{l}\tau^{m-1}v'(\tau)v(\tau) - \frac{m-1}{2}\tau^{m-2}v^2(\tau) - \frac{1}{m}\tau^m\left[v'(\tau)\right]^2 - \\ \frac{1}{2}(m-1)(m-2)Q_{m-3,0}\end{array}\right]. \quad \text{(A.88)}$$

For $m \leq 3$ we have

$$Q_{0,0} = \tau v^2(\tau) - \left[v'(\tau)\right]^2,$$

$$Q_{2,0} = \frac{1}{5}\left[2\tau v(\tau)v'(\tau) - v^2(\tau) - \tau^2\left[v'(\tau)\right]^2 + \tau^3 v^2(\tau)\right], \quad \text{(A.89)}$$

$$Q_{1,0} = v(\tau)v'(\tau) - \frac{\tau^2}{2}\left[v'(\tau)\right]^2 + \frac{\tau^2}{2}v^2(\tau) - \frac{3}{2}Q_{2,0}.$$

Integrals $Q_{m,1}$ containing the higher order derivative can also be defined via recurrent formulas:

$$Q_{m,1} = \frac{\tau^m}{2}v^2(\tau) - \frac{m}{2}Q_{m-1,1}, \quad \text{for } m \geq 1, \quad \text{(A.90)}$$

$$Q_{0,1} = \frac{1}{2}v^2(\tau).$$

We used a known feature of the Airy function, that is that a derivative of any order $n, n > 1$ of an Airy function $v(\tau)$ can be derived via polynomials of τ, function $v(\tau)$ and its first derivative $v'(\tau)$.

This concludes the calculation of the integrals of the Airy function products.

References

1 Fock, V.A. *Electromagnetic Diffraction and Propagation Problems*, Pergamon Press, Oxford, 1965.

2 Abramovitz, M. and Stegun, I. *Handbook of Mathematical Functions*, NBS, Applied Mathematics Series-55, Washington, 1964.

3 Bateman H. and Erdellyi A. *Tables of Integral Transforms*, McGraw-Hill, New York, 1954.

Index

Radio Wave. Alexander Kukushkin

ISBN: 3-527-40458-9